The
Global Structure
of
Visual Space

ADVANCED SERIES ON MATHEMATICAL PSYCHOLOGY

Series Editors: H. Colonius (*University of Oldenburg, Germany*)
E. N. Dzhafarov (*Purdue University, USA*)

Vol. 1: The Global Structure of Visual Space
by T. Indow

Advanced Series on Mathematical Psychology   Vol. 1

The Global Structure of Visual Space

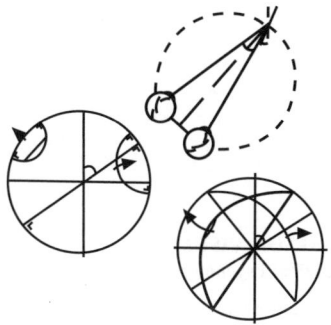

Tarow Indow
University of California, Irvine, USA

World Scientific

NEW JERSEY • LONDON • SINGAPORE • BEIJING • SHANGHAI • HONG KONG • TAIPEI • CHENNAI

Published by

World Scientific Publishing Co. Pte. Ltd.
5 Toh Tuck Link, Singapore 596224
USA office: Suite 202, 1060 Main Street, River Edge, NJ 07661
UK office: 57 Shelton Street, Covent Garden, London WC2H 9HE

British Library Cataloguing-in-Publication Data
A catalogue record for this book is available from the British Library.

THE GLOBAL STRUCTURE OF VISUAL SPACE
Advanced Series on Mathematical Psychology — Vol. 1

Copyright © 2004 by World Scientific Publishing Co. Pte. Ltd.

All rights reserved. This book, or parts thereof, may not be reproduced in any form or by any means, electronic or mechanical, including photocopying, recording or any information storage and retrieval system now known or to be invented, without written permission from the Publisher.

For photocopying of material in this volume, please pay a copying fee through the Copyright Clearance Center, Inc., 222 Rosewood Drive, Danvers, MA 01923, USA. In this case permission to photocopy is not required from the publisher.

ISBN 981-238-842-7

Foreword

In moving around, most animals guide the body in accordance with the information obtainable through vision. Humans are not exceptional. For humans, however, perceived space has played a more important role. Our ability to see the highly structured surrounds as they are has provided us with another ability to grasp the geometrical structure within a figure or between figures. This is a starting point of logical thinking in humans. As will be discussed in this book, "null or void" and "infinity" are beyond our direct perceptual experience. This may have something to do with the fact that humans have had difficulty in achieving these concepts. The use of the positional notation corresponding to 0 was an invention in India, perhaps around 6th century. Medieval abacus had the column corresponding to 0 but mathematicians had difficulty to conceptualize it. "The zero is something that must be there in order to say that nothing is there" (Menninger, 1969, p.400). Conceptualizing infinity is even subtler. Often people imagine a condition of "countless" as the image for infinity because this condition can occur as a phenomenon of our visual experience. The space we see extends in three directions. Gardner (1990) vividly describes how difficult it is to visualize something that is not in our perceived space. According to him, until the 19th century, mathematicians did not realize the possibility of extending Euclidean geometry to dimensions higher than three.

Investigating visual perception has been a main subject of experimental psychology. However, most studies have been concerned with local phenomena in the space we see. It is especially true since modern technology was introduced in experiments of visual perception. Generating visual patterns on a CRT by computer and the availability of fMRI (functional magnetic resonance imaging) have revolutionized the experimentation, and most investigators are more and more interested in detailed study of phenomena in a small area in the visual space. Consequently, studying the global structure inherent in what we see escapes the attention it deserves. This is the problem to be taken up in this book.

The publication of experimental results referred to in this book began in 1962. Many of my students at Keio University in Tokyo made contributions. The following names deserve explicit mention: Emiko

Inoue, Nobuko Momomi, Mitsuho Ohta, Mitsuko Shimada, Keiko Matsushima, Yasuo Nishikawa, Hisako Miyauchi, Noriko Yamashita, Akira Nakada, and Toshio Watanabe. I taught at Keio University until 1979, then moved to the University of California Irvine (UCI). Watanabe was accepted to UCI as a graduate student and received a Ph.D. with the study cited in Sec.5.1.1. Experiments numbered as 5 in Table 2.2 were performed by a student Kevin Wright.

While I was in Japan, I was given an opportunity to work as a research fellow at the Laboratory of Psychophysics in Harvard, 1963 - 1966. S.S. Stevens was the director. During that period, I met R.D. Luce, a professor of University of Pennsylvania at that time. Later, he invited me to the Institute for Advanced Study in Princeton as a visiting member, 1971-1972, and recommended me to UCI when he moved from UCI to Harvard. I am particularly indebted to these two distinguished scholars in the United States.

My study on visual space at UCI was partially supported by the National Science Foundation Grant, IST80-23893. As to the global structure of visual space, I have exchanged opinions with many scientists all over the world. Discussions with Patrick Suppes of Stanford University and Jan Drösler of University of Regensburg in Germany were especially helpful to me. The writing of this book is due to strong recommendation by two colleagues in the Society for Mathematical Psychology, E.N. Dzhafarov, a professor at Purdue University, and H. Colonius, a professor at Universität Oldenburg, Germany.

Irvine, California   Tarow Indow
December, 2003

Menninger, K *Number words and number symbols:A cultural history of numbers* (MIT Press, Cambridge, 1958).
Gardner, M The new ambidextrous universe: Symmetry and *asymmetry from mirror reflections to superstrings* (W.F.Freeman, New York, 1990).

CONTENTS

Foreword... v
Abbreviations and Symbols... ix

1. Visual Space.. 1
  1.1 Global Structure of Visual Space...................................... 1
      1.1.1 Features of VS.. 2
  1.2 Binocular Vision.. 8
      1.2.1 Cyclopean Vision in the Horizontal Plane of Eye-level 9
      1.2.2 3-D Cyclopean Vision... 11
      1.2.3 Spatial Behavior... 14

2. Luneburg Model.. 17
  2.1 P- and D-alleys, H-curves in the Horizontal Plane............. 17
      2.1.1 Experiments with Stationary Points......................... 17
      2.1.2 Discrepancy between $\{Q_i\}_P$ and $\{Q_i\}_D$................. 20
  2.2 VS as a Riemannian Space of Constant Curvature............. 22
      2.2.1 Riemannian Space of Constant Curvature............... 22
      2.2.2 Eudlidean Map (EM).. 26
      2.2.3 Equations of P- and D-alleys, H-curves in EM^2........ 29
  2.3 Theoretical Curves in X^2.. 33
      2.3.1 Luneburg's Mapping Functions................................ 33
      2.3.2 Equations of P- and D-alleys and H-curves in X^2..... 37
      2.3.3 Comments on Results of Alley Experiments............ 39
      2.3.4 Comments on Values of $-K$ and σ....................... 44
  2.4 Derivations and Explanations... 47
      2.4.1 Supplementary Explanations to Sec.2.2.1................. 47
      2.4.2 Derivations of Equations in Secs.2.2.2 and 2.2.3....... 51

3. Two Extensions of Luneburg Model................................. 67
  3.1 Alleys on a Frontoparallel Plane....................................... 68
      3.1.1 Theoretical Equations.. 68
      3.1.2 Experimental Results.. 75
  3.2 Direct Mapping according to Riemannian Metric............. 79
      3.2.1 Multidimensional Mapping according to Riemannian
           Metric.. 80
      3.2.2 Experimental Results.. 91

3.2.3 Concluding Remarks to Sec.3.2.................97

4. Visual Space under Natural Conditions...........103
4.1 The Perceived Sky and Ground........................103
    4.1.1 Bisection of the Sky.....................................104
    4.1.2 The Moon Illusion.......................................109
    4.1.3 Multidimensional Construction of the Night Sky........113
    4.1.4 Horizon..117
4.2 Scaling of Radial Distance δ_0.............................122
    4.2.1 Scaling based on Difference Judgment........122
    4.2.2 Scaling based on Ratio Judgment................129
    4.2.3 Discussion on the Form of $d(x)$.................133
4.3 Perceived Spatial Layouts under Full Cue Conditions........137
    4.3.1 Three Experiments......................................137
    4.3.2 General Discussion.....................................143
    4.3.3 Regular Triangles with S at the Barycenter........145

5. Related Experiments and Theoretical Considerations........149
5.1 Spatial Layouts in Frameless Visual Space........149
    5.1.1 An Experiment with Circles........................149
    5.1.2 Experiments using Triangles......................152
    5.1.3 General Comments and Derivations...........159
5.2 Mapping Functions...163
    5.2.1 Experimental Data.......................................163
    5.2.2 Theoretical Considerations.........................167
    5.2.3 Roles of Mapping Functions.......................173
5.3 Experimental Tests of Properties of VS as an R........175
    5.3.1 Two Experiments of Foley..........................175
    5.3.2 Bottom-up Experimental Approaches........180
5.4 Discussion on the Postulate that VS is an R........186
    5.4.1 Helmholtz-Lie Problem...............................186
    5.4.2 Congruence and Similarity in VS.................189
    5.4.3 Linear Perspective.......................................193
    5.4.4 Concluding Remarks...................................197

References...201
Author and Subject Indexes....................................211
Credits..216

Abbreviations and Symbols

S	subject (observer) in experiment
VS	visual space
X	physical space
E	Euclidean space
R	Riemannian space of constant curvature
K	Gaussian (totoal) curvature
r	curvature radius
EM	Euclidean map to represent R in E (Poncaré)
BS, BC	basic sphere, basic circle in EM
Σ in Sec.2.4.2	Klein's model for R^2 of $K < 0$
Ω in Sec.2.4.2	sphere representing R^2 of $K > 0$
Q, Q(x,y,z), Q(γ,ϕ,θ)	stimulus point in X, see Fig.2.3
γ, ϕ, θ	convergence, lateral, elevation angles
$P_V(\xi_V,\eta_V,\zeta_V)$, $P_V(\delta_0, \varphi_V, \vartheta_V)$	perceived point in VS, see Fig.2.3
$P(\xi,\eta,\zeta)$, $P(\rho_0, \varphi, \vartheta)$	point representing P_V in EM, see Fig.2.3
O or 0	origin of coordinates, body in X, self in VS, EM
DP(θ)	X^2 extending in depth along x with an elevation angle θ or its conterpart in VS and EM
horizontal DP	DP with $\theta = 0$
HP	frontoparallel plane in VS or its conterpart in X, EM
e, e_0	distance, radial distance (Euclidean) in X
ρ, ρ_0	distance, radial distance (Euclidean) in EM
δ, δ_0	distance, radial distace (perceptual) in VS, latent variable
d, d_0	scaled value of δ, δ_0

q	a constant defined in Eq.2.2.2 $= 1/2r$, $\sqrt{K}/2$ when $K > 0$ $\sqrt{-K}/2$ when $K < 0$
G	see Eq.2.2.1, Eq.3.1.1
σ	a parameter in the Luneburg's mapping function (Eq.2.3.1)
P-alley	parallel alley in X or its conterpart in EM
D-alley	(equi) distance alley in X or its counterpart in EM
H-curve	frontoparallel curve in X or its counterpart in EM
Hv	vertical frontoparallel curve
Ph, Dh	P-alley, D-alley in the horizontal direction on HP
MDS	multidimensional scaling
RMS	root-mean-square
EMMDS	see p.87
DMRPD	see p.90
c	a constant in EMMDS
sft, sm	subjective foot, meter, unit, Sec.4.2.1
( )	matrix
{ }	configuration of points
bold symbol	vector or matrix

1. Visual Space

1.1. Global Structure of Visual Space

Unlike sensory experience in other sense modalities, what one sees is a spatially structured scene. In a room, one sees furnitures in front of the wall and a part of garden through a window. Outdoors one sees trees and houses under the sky. Let us call this perception *visual space* (VS). Perception is due to a series of processes that starts from sensory stimulation and ends as the brain excitation. Taste is due to chemical stimulation to the tongue. As shown in Fig.1.1, VS is the end-product of the especially long series of processes, physical, physiological, and psychophysical. The first process is the formation of retinal images of the physical space on both eyes. This is a physical process and images are formed according to laws of optics. A physical object reflecting light is called a *distant stimulus* and its image on the retina is called a *proximal stimulus*. The retina is the light sensitive, innermost, nervous tunic of the eye. Photoreceptors in the retina are stimulated by the proximal stimulus and the excitation is conveyed to the visual cortex in the brain. This process is physiological. Due to the physiological excitation in the brain, one perceives a highly structured VS that consists of segregated objects (percepts), the background and the perceived self. This last process is psychophysical and governed by its own laws.

Most studies of visual perception are concerned with local phenomena in VS. However, the main concern in this book is the global structure of VS, and features of VS, topics to be discussed in this book, are enumerated in the next section. Abbreviations and symbols are defined when these appear for the first time and are listed before this chapter. For instance, the physical space is denoted by X to distinguish it from VS. Whenever it is necessary to make explicit the dimensionality of X

2   *Global Structure of Visual Space*

or VS, it is denoted by m and attached as the superscript such as X^m, or VS^2 (a perceived plane).

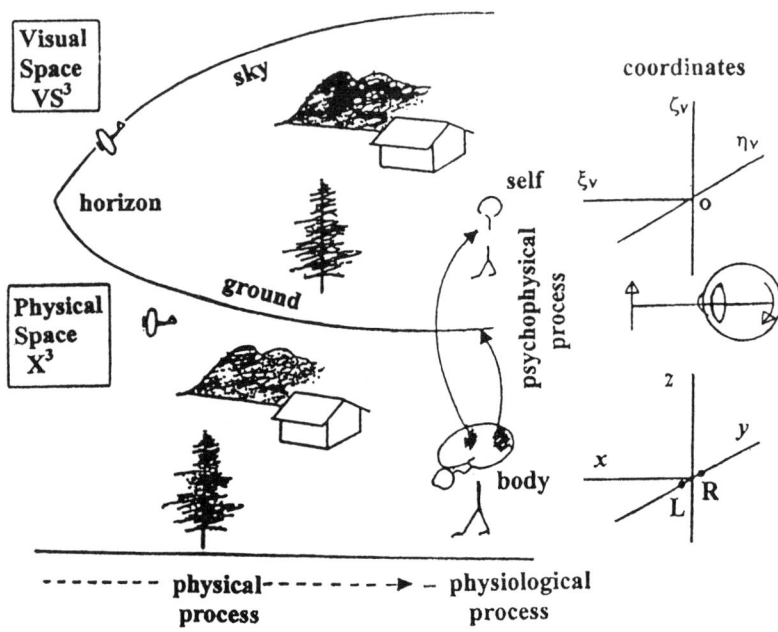

Fig.1.1 Physical space X^3 (natural scene) and visual space VS^3

1.1.1. *Features of VS*

VS1. Under the ordinary condition, VS^3 extends around the self in three directions. Cartesian coordinates of X^3 and VS^3 will be respectively denoted as x, y, z and ξ_v, η_v, ζ_v (Fig.1.1). The perceiver as a physical entity in X is called *body* whereas the perceived body in VS is called *self*. The self is mainly due to proprioceptive stimulation in the body. Excitation for visual percepts and excitation for the self occur in different parts of the brain, but the center of self plays the role of origin (O or 0) for ξ_v, η_v, ζ_v. When one says that "an object appears above and in the right direction", it means that the percept is localized in VS in that way with regard to the self.

Once, the following argument was made. An object forms its image on the retina and the retina is within myself. Nevertheless, I see the object in front of me. Furthermore, as an image in a camera, the retinal image of a tree is inverted up and down as well as left and right (Fig.1.1). Nevertheless we see the tree standing upright on the ground. Hence, there must be an appropriate "projection process" to make possible this inversion. To my best knowledge, Köhler (1929) was the first to point out that this argument stems from the confusion between self and body. The retina is inside the body in X and the spatial relation between the retinal image and the object is inverted in X. From this proximal stimulation, such a VS is generated in which the spatial relation between the percept and self matches the spatial relation between the distant stimulus (not the proximal stimulus) and the body in X. The real question is, not the projection, but how VS is generated in this way as the end-product of the series of processes.

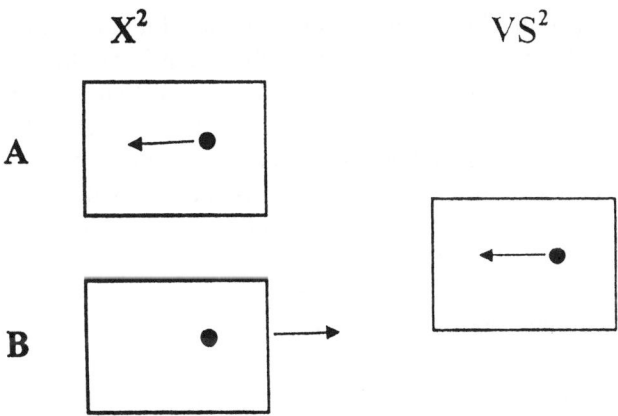

Fig.1.2 Induced movement

Suppose that a stimulus pattern, a rectangle with a dot inside, is presented (Fig.1.2). If the dot moves to the left (A), the observer perceives the movement in that way in the rectangle that stands still. The same movement is perceived when the rectangle is moved to the right (B), provided that this stimulus pattern is observed from a certain distance. This phenomenon is called induced movement and was studied carefully by Dunker (1929). The necessary condition for the dot alone to appear to move is that the physical movement of the rectangle with regard to the observer's body is small enough so that its movement is not peceptible. We are more sensitive to the change of relationship between dot and rectangle as its framework than the change between the rectangle and the body. Dunker reported a case that if the observer is close to the stimulus pattern and fixates on the dot, sometimes the observer feels to move to the left with the dot. We can have the same experience when we observe a stream from a bridge. We feel as if we are moving upstream. When spatial relationship between an object and its framework is changed, the change appears as perceptual movement of the object. This is an example of psychophysical laws that are active in the last process in Fig.1.1. Also, this shows that the self is a percept as the perceived dot is. A physicist studying snow described a queer experience he had when observing falling snow at night. He did not see in a beam of flashlight anything but a falling stream of snow, and he felt as if he were ascending to the sky.

VS2. Percepts in VS, including the self, are hierarchically related to each other. We say "a book is on the table", but it would not be natural to say "the table is beneath a book". As demonstrated by the induced movement (Fig.1.2), the dot is localized with respect to the rectangle acting as its framework. When the physical movement of rectangle to the right in B is beyond the threshold, the observer sees the dot moving to the left in the rectangle moving to the right. To the dot, the rectangle is the framework and to the rectangle, its surround or the self is the framework. The self is not the ultimate framework for other percepts in VS. In a room, the self is a percept localized with regard to the perceived floor and wall. Hence, it is possible to cause induced movement of the self to the left by moving the floor and wall to the right for the body staying still in the air. Outdoor, the ground and the sky are the largest possible framework in VS. When no framework is visible during a flight, the pilot may lose orientation.

VS3. VS is closed. At the end of the line of sight, there is always some percept, an object or background. Any distant stimulus in X, however far away, appears as a percept at a finite distance from the self and we never see anything to be at infinity, which means that VS has the boundary in all directions. This is the reason that the sky is perceived as a vault. At the boundary, we see something that has color, chromatic or achromatic. There is no percept appearing to be vacant, which means that the boundary of VS is continuous in the sense of having no gap. Namely, one cannot perceive "infinity" or "void". Vacant space lies between the self and the percept in the direcion of sight. This *terra incognita* is not a percept in the sense that one sees the backgound in between perceived objects on the boundary. The real vacancy is the backside of the self in VS. Physical objects far enough from the body in X, such as a flying plane or stars, are perceived as being embedded in the boundary (Fig.1.1). In the beginning, the plane is perceived to fly away under the sky but eventually the plane or its vapour trail starts to go down along the vault and finally vanishes to the horizon, if the horizon is visible. The horizon always appears at the height of the eye in VS. Why it appears in this way will be discussed in Sec. 4.1.4.

VS4. In the neighborhood of the self, VS is *veridical* to X, which means that the structure of VS coincides with that of X. The two are isomorphic and even "isometric" in a sense. This is the reason we can move around in X being guided by what we see in VS. To have a structured VS having this veridicality must have been far more advantageous in avoiding danger or in attacking prey compared with creatures that are only able to distinguish whether they are in the shade or not. The global coincidence of structure between X and VS holds only in the neighborhood of the self. As mentioned in VS3, the perceived distance from the self to a percept in VS beyond this range is not "proportional" to the distance between that object and the body in X. However, the "distorted" correspondence between VS and X in this region of VS does not affect directly our behavior on the ground. It may matter for a flying pilot. Fieandt (1966) described the trouble that aerial observers in antiaircraft batteries during World War II experienced. An enemy plane passing horizontally looked as if its course were curved.

VS5. VS is dynamic. If the whole area of retina is exporsed to homogeneous light of a sufficiently low intensity, one feels as if being surrounded by fog of light. This is called the Ganzfeld experience. The first experiment was performed in the psychological laboratory in Berlin University (Metzger, 1930). The subject was seated at 1.25 m from the wall of 6 m high. The wall covering the entire visual field of the subject was uniformly illuminated from far behind and above the subject (Engel, 1930). When the illumination was low, the subject felt like he was "swimming in the mist of light which extends from indeterminate distance". When the illumination was slightly increased, the subject felt as if being "inside of a balloon" and the curved surface appeared to be at about 50 cm in front of the self. When the illumination intensity is further increased, the curved surface was condensed to a vertical flat surface. This perceived surface appeared closer to the self and much brighter than when a wall of 1.25 m distance was observed under the same level of illumination in the ordinary condition. Metzger ascribed the change of appearance from the mist of light to the perceived surface to the heterogeneity in the light reflected from the physical wall. When the illumination was increased, micro-irregularity on the surface of wall introduced the heterogeneity in the proximal stimulus. Under the ordinary condition, retinal image consists of optical patterns and this is the necessary condition to generate a VS in the neighborhood of the self that is isomorphic to the structure of X. Still, as mentioned in the last part of VS3 and will be discussed later, how far VS extends from the self very much depends upon the total configuration of distant stimuli in X.

VS6. VS is stable in spite of the fact that it is based on multiple glances. Unless we fixate on an object, VS is a product based on multiple glances of X. Nevertheless, we see a coherent VS. When the line of sight is shifted from an object to another in X by moving the eyes or the head, the retinal image of X undergoes a continuous change. Nevertheless, we see a stable VS. We feel the change of line of sight within the VS. We take it for granted that everyone has a VS of the same structure when one faces the same physical conditions in X. Otherwise, we cannot hold conversation and all novels presuppose this unanimousness when describing a scene. In this article, no discussion will be made on the physiological mechanism that makes it possible for us to have this stable VS. The discussion will be focused upon the structure of VS that is almost independent of the retinal condition at a glance.

VS7. When we observe the rectangle A in Fig.1.2, it is perceived as a rectangular form not only when the sheet of paper is perpendicular to the line of sight. Even when the sheet is tilted, the percept will still be a rectangle in spite of that the retinal image is not rectangular but trapezoidal under this condition. The end-product from the trapezoidal retinal image is the rectangle tilted in VS. This is called *shape constancy*. When we hold a pencil at a distance of 25 cm and then move it to 50 cm distance, the length of its retinal image is reduced to a half. Nevertheless, we see the same pencil held at different distances from the self. The change of distance is noticed but the perceived length of the pencil remains the same. This is called *size constancy*. In general, the pattern we see in VS^3 is closer to the pattern of object in X^3 than to the pattern of its 2-D retinal image, which is helpful for us in indentifying an object in X. If the final product of the processes is not stabilized in this way, VS would be very messy. Drösler (1995, p.81) described his "nauseating" experience when he saw a TV picture from a camera inside the helmet of one of the players of American football game. Under this condition, constancy fails to work and the impression was that "teams of dwarfs" were playing football "inside a running washing machine". This book will be concerned with the feature of VS that is not directly bound to the retinal image.

VS8. VS is geometrical. One perceives geometrical patterns in VS; curves, straight lines, intersections, betweenness, distances, angles, triangles, and so forth. Perceived length, angle, area are magnitude-like. Some are large and some are small. Furthermore, humans can recognize congruence and similarity between figures. Congruence means that two figures are geometrically the same in all the respects. Similarity means that two are of the same form but differ in size. Because we can recognize similarity, we can identify objects in a picture that are much smaller than the percepts in the original scene. This problem will be discussed in Sec.5.4.2. Perhaps, on the basis of perceived patterns in VS, human being came to obtain the idea of geometry and also logical thinking (Indow 1999).

Perception of geometrical patterns and the geometry of VS itself are the main problems to be discussed in this book. It is of importance to keep in mind that VS is not a solid container in which various percepts are placed. What is meant by the geometry of VS is not the geometry of this solid container. It refers to the geometry according to which the perception of a geometrical pattern is structured.

8 *Global Structure of Visual Space*

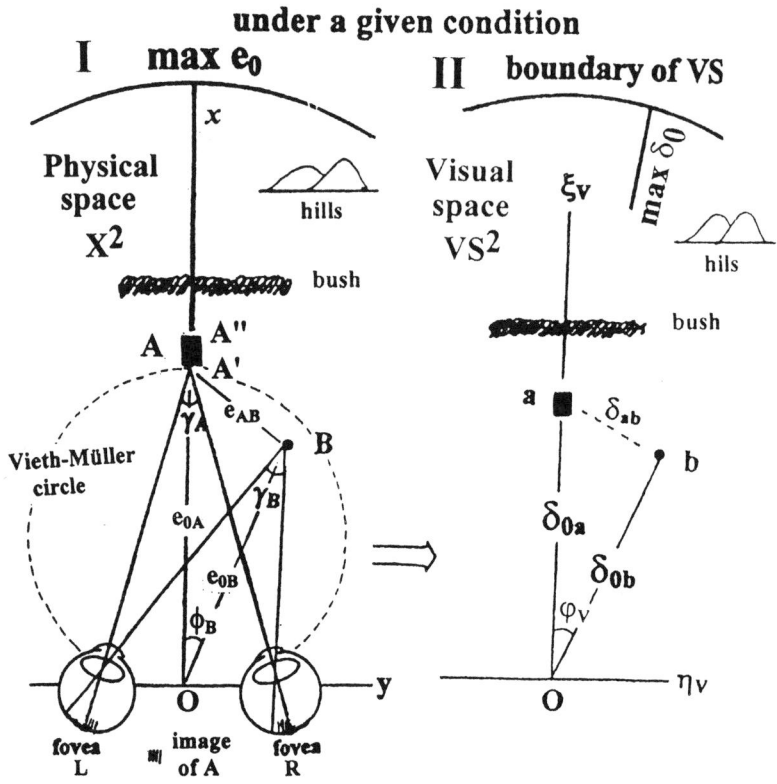

Fig.1.3 Convergence and corresponding points

1.2. Binocular Vision

Human beings have two eyes both facing forward. Hence, the two eyes have overlapping images of the X. A united VS is generated from retinal images of the left and right eyes. Often, this is called *cyclopean vision* after the mythical creature who looked out on the world through a single eye in the middle of his forehead. When we fixate on an object, its image falls on the fovea of the retina of each eye. There are two kinds of photosensitive cells in the retina, cones and rods. Color vision is due to activities of cones, and rods convey achromatic information only. The fovea consists of the cones and is most highly adapted to yield high visual acuity.

1.2.1. *Cyclopean Vision in the Horizontal Plane of Eye-level*

Let us begin with the simplest case. Suppose that A and B in Fig.1.3I are two thin objects in the horizontal plane X^2 of eye-level for a subject looking straight ahead. A plane extending from both eyes in the depth direction will be denoted as DP in this book. The horizontal plane is the DP parallel to the ground. The Cartesian (rectangular) coordinate axis *y* is the line passing through the nodal points of the eyes, and the other axis *x* starts from the middle point O. Correspondingly we can introduce into VS^2 the Cartesian coordinate axes. The axes are denoted as ξ_V and η_V. The origin O is the center of the self from which the direction of VS is defined (Fig.1.3II). It is not easy, perhaps not too important, to precisely localize O within the self. Drösler (1966) called it "Ich-mitte".

Suppose that the subject fixates on the closer corner A' of a bar A in X^2. The angle γ_A is called *convergence angle* (Fig.1.3I). Through the mechnism called accommodation, the lenses of two eyes are automatically adjusted so as to cast the sharppest images of the corner A' at the foveae of two eyes. The same location from the fovea along the equator in each retina is called corresponding point. The dotted circle passing through the nodal points of both eyes and the fixated point is called the *Vieth-Müller circle* and any point on this circle is supposed to project its image on the corresponding points of the two retinae. The other corner A" of the bar A is not on the circle, and the image of A on the right retina extends from the fovea to the left, whereas that on the left eye extends from the fovea to the right. The angular difference between the two retinal positions of A" is called *binocular disparity*. Due to this disparity, the bar A appears in VS as a percept "a" extending in the depth direction (Fig.1.3II).

There is a point object B in Fig.1.3A. The lateral angle is denoted as ϕ; $\phi_A = 0$ and $\phi_B > 0$ in this direction. If B is on the Vieth-Müller circle, it forms its images on the corresponding points and the images are sharpest. We will see a point "b" in the direction ϕ_V that corresponds to ϕ_B. In Fig.1.3I, B is not on the Vieth-Müller circle for the eyes fixating

10 *Global Structure of Visual Space*

A. When the binocular disparity is too large, we see two separate percepts of B in VS. If two images are within the so-called Panum's fusional area, we have one perceived point b in VS. According to whether B is inside or outside the Vieth-Müller circle, b is closer or farther away than a. Stereoscope is an instrument to produce depth impression between objects by the use of binocular disparity.

In X^2 in Fig.1.3I, two radial distances from O are denoted as e_{0A} and e_{0B}. The corresponding perceptual distance form the self are denoted as δ_{0a} and δ_{0b}. What is implied in VS4 in 1.1 is that, in this neighborhood of the self, a appears on the axis ξ_V ($\varphi_V = 0$) and $\varphi_V > 0$ for any object such as B in which $\phi > 0$. Furthermore, the ratio δ_{0a}/δ_{0b} is equal or close to the ratio e_{0A}/e_{0B}. The former ratio is between two perceived distances. For a perceived distance, δ_0 in Fig.1.3II, we have no yardstick to measure it. Nevertheless, we can feel the ratio if two δ's are not too different. This problem will be discussed in Sec.3.2.1.

For A and B in Fig.1.3I, perception of δ_0 is not directly determined by e_0. The perceptual size of the bar in the η_V-direction is determined by the size of retinal image of the bar, but there is no retinal image of e_{0A} itself. The perception of δ_{0a} is mediated by distance cues, accommodation and convergence that covary with e_{0A}. The range of e_{0A} in which these cues are effective is limited. Accommodation of lens may play some role in determining δ_{0a} only if e_{0A} is less than 1.5 m, and convergence remains practically constant if e_{0A} exceeds 15 m. Suppose that bush and hills are visible far behind A and B in Fig.1.3I. Let us denote by e_0 the radial distance from the body to an object in general. The bush and hills will appear in VS at radial distances δ_0's with lateral angles φ_V's (Fig.1.3II). Insofar as φ_V is concerned, its relation with ϕ may not be so distorted, and the ratio (φ_V to the hills) over (φ_V to the bush) may be equal to the ratio (ϕ_V to the hills) over (ϕ_V to the bush). However, this proportionality does not hold between δ_0 and e_0 for objects beyond a certain range. The ratio (δ_0 to the hills) over (δ_0 to the bush) may be smaller than the ratio (e_0 to the hills) over (e_0 to the bush). How (δ_0 to a percept) is related to (e_0 to that object) highly depends upon what are visible in VS. Eventually, even in an unobstructed field of vision, there is the boundary of VS in a given direction at which δ_0 reaches its maximum, max δ_0 (VS3 in 1.1). Correspondingly, max $e_0(\phi)$ exists in X in the sense that all objects beyond this distance in this direction will appear at max $\delta_0(\varphi_V)$. The boundary of VS may not be visible by itself, but it exists functionally.

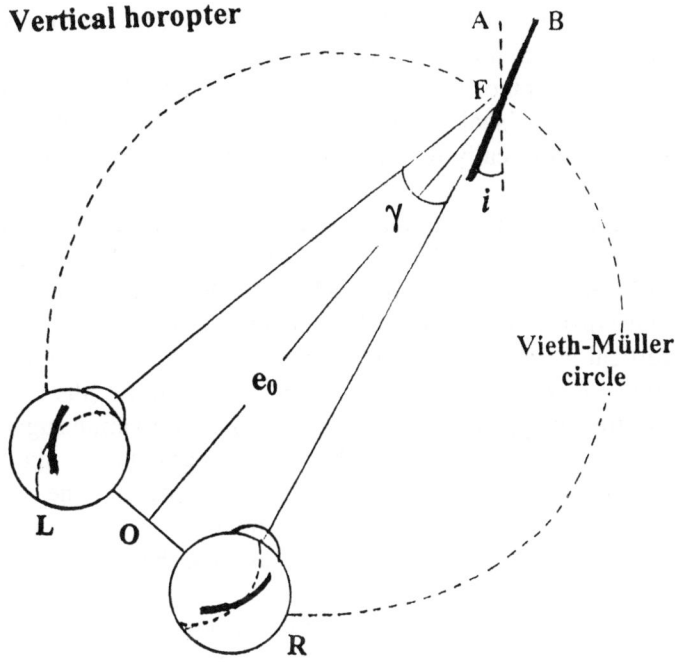

Fig.1.4 Vertical horopter

1.2.2. *3-D Cyclopean Vision*

The Vieth-Müller circle in Fig.1.3I is often called horizontal *horopter*. The horopter is the locus of points in X that project images to corresponding points in each retina. If "correspondence" is assumed to hold between any pair of points that have geometrically the same retinal eccentricity from the respective foveae, points on the Vieth-Müller circle stimulate corresponding cells on the equator passing through the fovea in each eye. According to this definition of corresponding points, the *vertical* horopter for distant stimulus points that are off from the plane X^2 in Fig.1.3I is the straight line perpendicular to the Vieth-Müller circle at the fixation point. The dotted straight line A passing F in Fig.1.4 projects its image on the dotted meridian curves in the retina in Fig.1.4. However, corresponding points can be understood in different ways. Ogle (1964) listed its five definitions such as (3) the positions of the points will lie in the "center" of the region of binocular single vision and (4) the positions of the points will be such that the stereoscopic sensitivity to changes in position will be a maximum". According to Tyler (1991), "corresponding points define zero binocular disparity, and are considered to project to the same region of visual cortex. Thus, corresponding points and hence a horopter are empirically determined by various methods according to its definition.

Some evidence has been found to show that the vertical horopter is not represented by the vertical dotted straight line A in Fig.1.4 (Amigo 1974, Cogan 1979, Krekling and Blika 1983, Ogle 1964, Nakayama 1977). If no binucular disparity exits for all points on the bar A, it is expected to appear vertical in VS^3. However, the bar A appears inclined toward the self and when it is inclined around F by an angle i in X^3 (the thick bar B in Fig.1.4), then it appears vertical in VS^3. This observation suggests that corresponding points are not on the dotted meridians of the retinae. Suppose that the thick curves in Fig.1.4 represent the series of corresponding points (SCP). We can estimate how SCP is tilted with regard to the meridian in the retina by measuring the angle i or by other methods.

Fig.1.5 represents the eyes of a person standing at G on the ground and fixating at F on the x-axis. Let the inter-pupil distance a be 6.4 cm and the height of the person be such that O is h cm high from the ground. The distance OF is denoted as e_0. Suppose that the series of corresponding points SCP (dotted curve) is tilted from the meridian (contimuous curve) by an angle of $\varepsilon/2$ in each retina. Consider two planes passing F, one passing L and the other passing R, that meet at O' on z-axis with the angle of ε. Then, all points on the plane O'LF form images on SCP of the left retina and all points on the plane O'RF form images on SCP of the right retina. Hence, the intersecting line O'F is the vertical horopter and it has the inclined angle i with regard to the z-axis. Then, the angle i is related to ε, e_0, and a as follows. The distance OO' is denoted as Z_0.

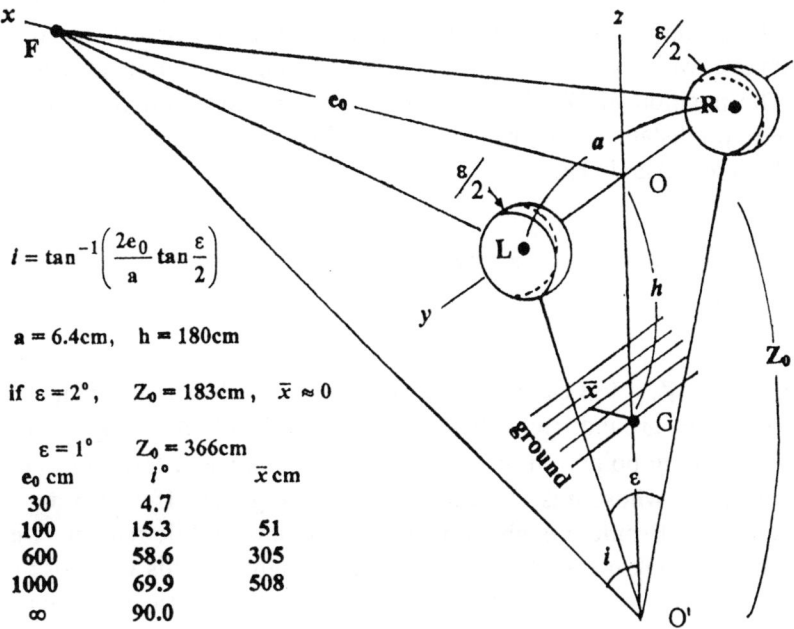

$$i = \tan^{-1}\left(\frac{2e_0}{a} \tan\frac{\varepsilon}{2}\right)$$

a = 6.4cm, h = 180cm

if $\varepsilon = 2°$, $Z_0 = 183$cm, $\bar{x} \approx 0$

$\varepsilon = 1°$ $Z_0 = 366$cm

e_0 cm	$i°$	$\bar{x}$ cm
30	4.7	
100	15.3	51
600	58.6	305
1000	69.9	508
∞	90.0	

Fig.1.5 Inclined vertical horopter

$$\tan i = \frac{e_0}{Z_0}, \quad \text{from the right triangle OO'F,}$$

$$\frac{a}{2} = Z_0 \tan\frac{\varepsilon}{2}, \quad \text{from the right triangle OO'L or OO'R}$$

and hence

$$i = \tan^{-1}\left(\frac{2e_0}{a} \tan\frac{\varepsilon}{2}\right)$$

Most experiments on the vertical horopter suggest ε to be 1° or 2° for human. Helmholtz pointed out that the inclined horopter due to this tilt of SCP is convenient for walking (Kahl 1971). If the vertical horopter is the straight line A in Fig.1.4 ($\varepsilon = 0$), all points on the ground have binocular disparities. In contrast, if ε is around these values, some points on the ground in the direction to walk can form their images on SCP of each eye, which will be helpful in detecting roughness of the ground.

The position of O', Z_0, changes according to ε and e_0 for a fixed interpupil distance a. Suppose that $a = 6.4$ cm. Let $\bar{x}$ be the intersection of the ground and O'F.

$$Z_0 = \frac{3.2}{\tan\frac{\varepsilon}{2}}, \qquad \bar{x} = (Z_0 - h)\tan i$$

For $\varepsilon = 2°$, $Z_0 = 183$ cm, and a person whose h is close to this value, $\bar{x} = 0$ and the ground right below the feet forms its image on the corresponding points. For $\varepsilon = 1°$, $Z_0 = 366$ cm, and a few values of $\bar{x}$ for a person whose h is 180 cm are given in Fig. 1.5. If $e_0 = 6$ m, $\bar{x}$ is about 3 m. Beside, i is about $5°$ for $e_0 = 30$ cm. While reading, we do not hold a book perpendicular to the line of sight and this value of i may be close to the preferred inclination. The geometrical relationship between horopters and SCP discussed above is schematic, but suggestive. Cooper and Pettigrew (1979) showed by simultaneously mapping the receptive field positions of binocular cortical neurons that the SCP is tilted in cat and owl so as to yield the inclined vertical horopter that is meaningful for their behavior.

1.2.3. *Spatial Behavior*

It is of vital importance for the survival of animals and human to have appropriate and correct information on X. Except for a reflex movement to avoid something suddenly flying toward the body, human moves the body in accordance with the spatial relationship between a percept and the self. It is because our cyclopean vision yields highly structured VS around the self and VS is veridical to X in the region where we move around (VS4 in 1.1). Animals of higher level exhibit behavior suggesting that they have VS as we do. However, to be able to behave appropriately in X does not nessessarily mean that the animal has a structured VS like ours. An experiment on spatial vision of chameleon was reported by Harkness (1977). This study has been widely cited (*e.g.*, Collet and Harkness 1982, Howards and Roger 1995).

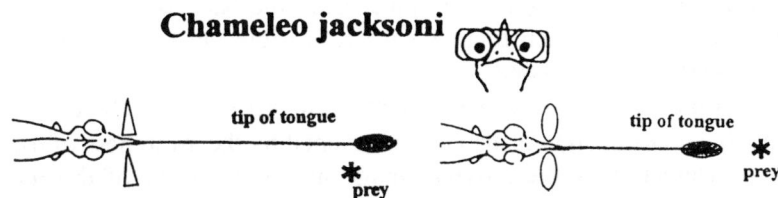

Fig.1.6 Chameleon shoots out the tongue to a prey

Chameleon has terretlike eyes, each moving independently to scan a very wide range of X on each side of the body in searching prey. Once a prey is detected, both eyes are swivelled directly towards the prey, and the chameleon shoots its tongue out like a flash at the prey in the range of 25 cm. Let object A in Fig.1.3I be a prey, then chameleon must have the precise information on e_{0A}. Chameleon has large eyes and they have well-developed fovea in each retina. However, it does not necessarily mean that the chameleon has the awareness of δ_{0a} by cyclopean vision and stretches the tongue in accordance with δ_{0a}. First, it was proved that the targeting is not based on the olfactory cue and it is visually controlled. When a lateral prism was placed in front of each eye that bends the ray of light, the direction of the tongue shooting was shifted as calculated from the prism (the left of Fig.1.6). When a lense was placed in front of each eye, the distance of tongue stretch was altered as calculated from the lense (the right of Fig.1.6). The distance information is not based on the convergence angle γ_A in Fig.1.3I. When one eye was blocked, the coincidence between the tongue stretch and e_{0A} was not affected. It was observed that the chameleon maintained the pupil dilation while aiming. Then, the veridical tongue shooting must be regulated by the monocular focusing mechanism, accommodation. By an auto-focusing mechnism, the lense is adjusted until the sharpest image of the prey is obtained and at the same time, this condition of accommodation directly dertermines the movement of the tongue. If this is the case, the chameleon does not need to have size constancy and shape constancy (VS7 in 1.1). Retinal image of a prey changes its size and shape according to e_{0A} and the viewing angle. In order to have the veridical information on e_{0A} from these cues, the chameleon must have the information on the size and shape of the prey in advance. If the tongue shooting is regulated by focusing mechnism, what the chameleon needs to do is to distinguish an object from the backgound, and if it is identified as a prey, then focus on it. It is irrelevant how shape of the prey appears when focussed.

Our reflex to move the body in order to avoid some object A that suddenly comes in sight is also not intervened by the awareness of δ_{0a}. It may be triggered by the moving image in the periphery of the retina. However, most movements around the body are guided by VS. To grasp an object A is guided by δ_{0a}. If A is a moving object, the movement of hand is controlled by the perceived path of its percept in VS. VS around the self consists of percepts localized at various distances δ_0 and various lateral directions φ with various angles of elevation ϑ. For eyes fixating object A in Fig.1.3, all points in X^3 that are not on the Vieth-Müller circle or not on the vertical horopter form their images on the retinae that have binocular disparities. The disparity within the images of object A yields percept a of 3-D solid form and the angle γ_A determines the perceived radial distance δ_{0a}. The angle ϕ_B and the difference between γ_A and γ_B determine the position of b, δ_{ab}, and δ_{0b}. When the fixation moves to B, then δ_{0b} is determined by γ_B. Our VS^3 is generated in such a way that we do not feel any contradiction between δ_{0b} when A is fixated and δ_{0b} when B is faxated. Of course, we see VS^3 of the about the same strucuture when we close one eye. Basically, however, our VS^3 is generated as stated above. The situation may be the same for mammals with frontal eyes, such as cats.

Animals with laterally placed eyes such as birds have a limited overlapping binocular vision. Unlike chameleon's eyes, bird's eyes are not terretlike. Their VS may be generated in a different way from ours. One advantage of having laterally placed eyes is that the animal can have a wide view on both sides of the body. Our field of vision is rather limited when the eyes are fixed. The monocular field of human vision spanes about $37°$ on each side and the binocular field of vision covers about $115°$ in front of the body. However, through multiple glances of X^3, human has a unified VS^3 around the self (VS6 in 1.1.1).

2. Luneburg Model

It was pointed out in Chapter 1 that VS, as the product of cyclopean vision, contains geometrical percepts and VS itself may be of geometrical nature. The simplest case will be dealt with in this chapter. In Sec.1.2.3, some geometrical analysis was tried. However, it was trigonometry about the condition in the physical space X^3 including the retinas. Naturally, X^3 was regarded to be Euclidean. What is to be discussed from now on is concerned with geometry of percepts in VS or VS itself. Then, it is not *a priori* clear what geometry is most appropriate for this purpose. Luneburg (1947, 1950) was, perhaps, the first to have tried to answer this question, although what is dealt with is VS under very limited conditions. It is *frameless* VS in which no framework, *e.g.*, the wall or ground, is visible and all percepts are localized with regard to the self only.

Luneburg was a geometer. It seems to me that two main motives led him to this problem. One is the Ames' demonstrations, such as the distorted room (Sec.5.2.3), that were well known at that time. The other is findings in the so-called alley experiment. The findings are already known since the beginning of the twentieth century. This experiment is explained in the next section.

2.1. P-and D-alleys, H-curves in the Horizontal Plane

2.1.1. *Experiments with Stationary Points*

As in Fig.1.3I, all stimulus points Q_i are presented on the horizontal plane DP at the eye-level of the subject S. The same coordinates (x, y) will be used to denote positions of Q_i. In Fig.2.1, Q_i shown by crosses are fixed; the farthest pair $Q_1(x_1, \pm y_1)$ and $Q_\alpha(x_\alpha, 0)$ on the x-axis, $\alpha = A$,

18 *Global Structure of Visual Space*

B, ... E. Other points Q_i (black circles or white diamonds), $i = 2, 3, ..$, are movable and their positions are adjusted by the S (remote control) or by the experimenter according to the order of the S. The series of black dots are the results when their positions are adjusted so that the two series appear straight and parallel to each other. The series will be called *parallel alley* (P-alley). The series of white diamonds are the results when their positions are adjusted so that $Q(x_i, y_i)$ and $Q(x_i, -y_i)$ appears to be symmetric to the median line with the same lateral separation. Namely, the perceived distance δ_i between the two points is constant in VS. The series will be called (equi-)*distance alley* (D-alley). All the adjustments of position are along the direction of the *y*-axis.

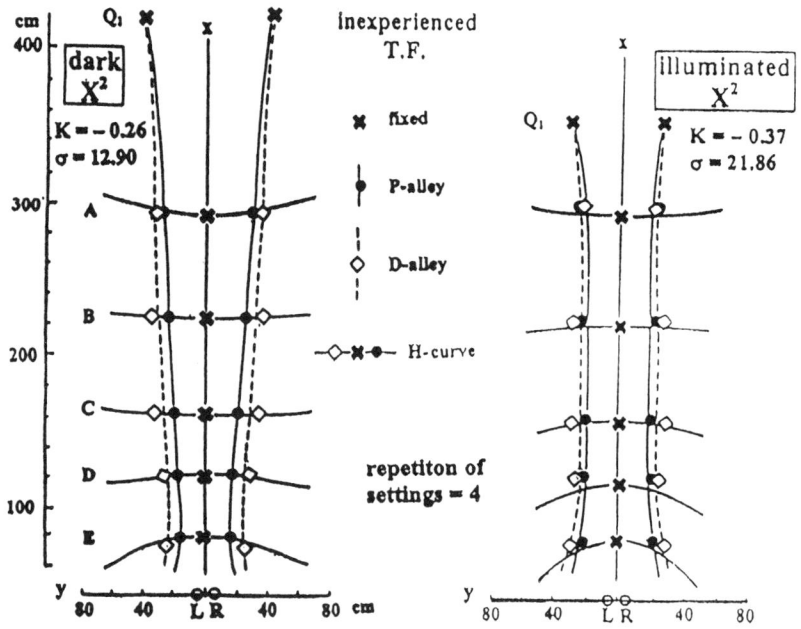

Fig.2.1 P-and D-alleys and H-curves in the horizontal DP

The two alleys are constructed separately. Namely, when a P-alley is being constructed, Q's represented by white diamonds are not presented. In this experiment, one more kind of adjustment is requested after P- and D-alleys have been obtained. Five Q's in a row from left to right are presented; $Q_\alpha(x_\alpha, 0)$ in the center (cross) and two black dots and two white diamonds on P- and D-alleys to the S. Then, the task of S is to adjust positions of black dots and white diamonds in the direction of the x-axis so that all the five Q's appear prontoparallel, flat from left to right. Each series, A to E, thus adjusted will be called a *frontoparallel curve*. This will be abbreviated as H-curve because the curve has been once called a horizontal horopter.

In Fig.2.1, L and R represent two eyes of the S. The left plot is an example when small light points were presented as Q's in a dark room. Light points, adjusted to appear with the same brightness, looked like fireflies. A girl observer once said "this is an experiment with fairies". The right plot is an example when small black balls (0.5 cm in diameter) were used as Q's. These were presented on the table covered by white sheets of paper. The table was illuminated from above (about 70 lux in the center). Since the eyes of S were at the level of the table surface, the S could not see edges of the table and the table was surrounded by white curtains. Namely, the S saw only black balls against the white background. In both experiments, the head of S was fixed but the S was encouraged to move the eyes to scan the series of Q's. Adjustments were repeated four times on separate sessions and mean positions $(x_i, +y_i)$ and $(x_i, -y_i)$ were further averaged to give the symmetric results. The final configuration thus determined will be denoted as $\{Q_i\}_P$, $\{Q_i\}_D$, and in the case of H-curve as $\{Q_i\}_H$. In practice, in order to avoid the occlusion of far Q's by closer Q's, the plane at which $\{Q_i\}$ were presented were slightly below the eyes of S. The S perceived all points on the horizontal DP extending forward from the self. In all experiments performed in my laboratory, results were always analyzed individually. Fig.2.1 shows $\{Q_i\}_P$, $\{Q_i\}_D$, and $\{Q_i\}_H$ of a S who participated for the first time in this kind of experiment (Indow, 1982). The theoretical curves fitted to $\{Q_i\}_P$, $\{Q_i\}_D$ and $\{Q_i\}_H$ are also shown. The explanation about these curves and two parameters, K and σ, will be given in Sec.2.2.

The two alleys do not coincide and $\{Q_i\}_D$ lies outside $\{Q_i\}_P$. This fact was sporadically noticed by Hillebrand (1902) and systematically studied by Blumenfeld (1913). Since then, Hardy, Rand, and Rittler (1951), Zajaczkowska(1956a, b), Squires(1956), Shipley(1957), Indow, Inoue and Matsushima, 1962a, b, 1963), *etc.* reported the same results. All suggest that "straight and parallel" does not mean "being equally separated" in this VS^2. However, before concluding in this way, it is necessary to show that the finding is not an artifact due to the experimental procedure or to the direction of eye movement during the observation.

2.1.2. Discrepancy between $\{Q_i\}_P$ and $\{Q_i\}_D$

The standard procedure of constructing P- and D-alleys is as follows. First, $Q_1(x_1, \pm y_1)$ and $Q_2(x_2, \pm y_2)$ are presented and $\pm y_2$ are adjusted. Then, $Q_3(x_3, \pm y_3)$ are presented. In the case of P-alley, $Q_1(x_1, \pm y_1)$ and $Q_2(x_2, \pm y_2)$ remain visible and the S sees three Q's on each side. After $\pm y_3$ have been adjusted, $Q_4(x_4, \pm y_4)$ are added. Namely, the S sees four Q's on each side while adjusting $\pm y_4$, and so on. When all the pairs have been adjusted, the S is allowed to readjust any Q_i, $i > 1$, if it is felt necessary, so that the two series appear to be straight and parallel. In the case of D-alley, $Q_2(x_2, \pm y_2)$ are turned off when $Q_3(x_3, \pm y_3)$ are presented and only two pairs, $Q_1(x_1, \pm y_1)$ and $Q_3(x_3, \pm y_3)$, remain visible. When $\pm y_3$ have been adjusted, $Q_3(x_3, \pm y_3)$ are turned off and $Q_4(x_4, \pm y_4)$ are presented. Namely, the S adjusts $\pm y_4$ so that the lateral separation of $Q_4(x_4, \pm y_4)$ appears to match that of $Q_1(x_1, \pm y_1)$, and so on. After all the pairs have been adjusted, all Q's are turned on to check that all appear to be equally separated. The S is allowed to readjust any pair, except $Q_1(x_1, \pm y_1)$, if it is felt necessary.

That $\{Q_i\}_D$ lies outside $\{Q_i\}_P$ is not because the S always adjusted $Q_i(x_i, \pm y_i)$ against $Q_1(x_1, \pm y_1)$ while constructing D-alley. The same discrepancy was observed even when D-alley was constructed in different ways. Once, when the S adjusted $Q_i(x_i, \pm y_i)$ to construct D-alley, all already adjusted pairs were left visible as in P-alley. Once, all the pairs $Q_i(x_i, \pm y_i)$ were simultaneously presented with random lateral separations, $i = 1, 2, \ldots$, and the S adjusted $\pm y_i$ until two series, one in the left and the other in the right of the x-axis, appear straight and parallel or two points appear equally separated along the y-axis. Always, $\{Q_i\}_D$ lay outside $\{Q_i\}_P$ (Indow and Watanabe, 1984a).

When constructing $\{Q_i\}_P$, it is natural for the S to scan the two series of Q's along the x-axis to see whether the two appear "straight and parallel". When constructing $\{Q_i\}_D$, the main direction of scanning is naturally along the y-axis to see whether each pair of Q's is "equally separated". Should this difference of scanning direction affect where Q's appear as points in VS, $\{Q_i\}_P$ and $\{Q_i\}_D$ will be different in X^2 even if two series of points are the same in VS^2. In order to test this possibility, the following experiments were performed (Indow and Watanabe, 1984a).

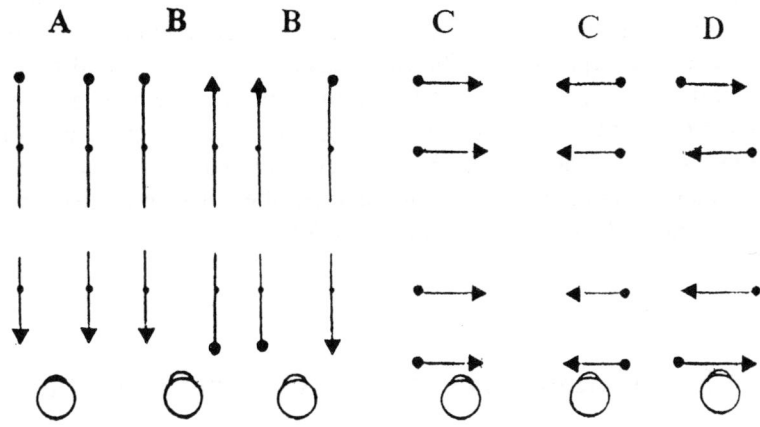

Fig.2.2 Alleys of moving points

In Fig.2.2, observation was made in two ways, dark and illuminated conditions. In the former, each Q_i was a light point that was turned on and off independently. In the latter, the S observed the field through a horizontal slit. Since the field was completely surrounded by black sheets of paper, what the S saw was the black background dimly illuminated (1.8 fL in the direction of S). Each Q_i was a white-painted metal rod that momentarily popped up into the field from below when an electric pulse was sent. Pairs $Q_i(x_i, \pm y_i)$ were presented one by one, $i = 1, 2,$ ..., with an appropriate time interval to generate an apparent movement between Q_i and Q_{i+1}. In Fig.2.2A, the S perceives two light points moving closer in the dark condition or forward movement of a pair of white rods in front of the black background in the illuminated condition. Presentation of Q's was controlled by an electronic device. Duration of $Q_i(x_i, \pm y_i)$ and the speed of the apparent movement were varied in two ways.

In Fig.2.2A, the S adjusted position y_i on the both sides until the movement appeared straight and parallel or with a constant lateral separation. Presentation of $Q_i(x_i, \pm y_i)$ was repeated until the S was finally satisfied with the settings. In Fig.2.2B, the direction of the movement was reversed between the two sides. Always, exactly the same discrepancy was found between $\{Q_i\}_P$ and $\{Q_i\}_D$. In C and D, two points in each pair were successively presented. In C, the S saw the entire series of Q_i move from left to right or right to left. In D, the direction of movement was alternated between Q_i and Q_{i+1}. Still the S was able to adjust $\pm y_i$ so that the apparent movement was either between two straight and parallel lines or with a constant separation. The same discrepancy was found between $\{Q_i\}_P$ and $\{Q_i\}_D$. In B and D, it was impossible for the S to pursue the movement with the eyes. Hence, that $\{Q_i\}_D$ lies outside $\{Q_i\}_P$ cannot be regarded as an artifact due to the difference of scanning direction. The discrepancy is genuine and "straight and parallel" does not mean "being equally separated" in the horizontal DP in VS.

2.2. VS as a Riemannian Space of Constant Curvature

Let us denote by $\{P_{Vi}\}$ the perceived configuration of points in VS when one observes a stimulus point configuration $\{Q_i\}$ in X. In the alley experiments being discussed in this chapter, the S adjusts $\{Q_i\}$ in X^2 so that $\{P_{Vi}\}$ satisfies the specified condition, and it was found that "straight and parallel" does not mean "being equally separated" in the horizontal DP. This finding implies that $\{P_{Vi}\}$ as the end-product (Fig.1.1) is not structured according to Euclidean geometry when it is formed in the DP. Luneburg proposed a hypothesis that $\{P_{Vi}\}$ in the DP is structured according to a Riemannian geometry. For simplicity's sake, this hypothesis is regarded as equivalent to say that this VS^2 is a Riemannian space.

2.2.1. Riemannian Space of Constant Curvature

Mathematically, many geometries and many spaces of different properties are known. Topology is concerned with spaces having no metric. Riemannian space in general is a metric space that is locally Euclidean. To say that VS is a metric space means that any length δ_{ij} we see in VS between two points P_{Vi} and P_{Vj} (the self is a point P_{Vo}, Fig.1.3) satisfies the Fréchet's conditions,

1. $\delta_{ij} \sim \delta_{ji}$, $\delta_{ij} > 0$, $\delta_{ii} \sim 0$ (symmetric and non-negative)

2. $\delta_{ij} \oplus \delta_{jk} > \delta_{ik}$ when three points are not collinear (triangular inequality)

3. $\delta_{ij} \oplus \delta_{jk} \sim \delta_{ik}$ when three points are collinear (additivity)

where " $\sim$, $>$, and $\oplus$" respectively mean "to appear equal to, longer than, and concatenated". Because δ's are latent variables that cannot be experienced by anybody other than the S, it is impossible to present the direct proof that δ's satisfy these conditions. We can ask the S to assess the ratio of two δ's and to obtain scaled values d that are supposed to be proportional to δ. As will be discussed in Sec.3.2.1, d's satisfy these conditions in the part of VS that is related to the alley experiment. In this book, let us assume that VS is a metric space when one perceives a structured pattern $\{P_{Vi}\}$, for the reason that we feel δ as a magnitude like percept and the above stated conditions are not contradictory to our perceptual experiences.

A metric space that is locally "Minkowskian" is called Finsler space. A metric space that is locally "Euclidian" is called Riemannian space. What is meant by "Minkowskian" or "Euclidian" will be explained in Sec.2.4. Let us assume that VS is a Riemannian space. It will be discussed in Sec.5.4 that this assumption is not contradictory to our visual experiences.

Riemannian geometry is an extension of the geometry of curved surfaces in a 3-dimensional Euclidean space. A parameter called the curvature is associated with each point in a Riemannian space. The curvature can be defined in various ways. In this book, only *Gaussian total curvature* K will be used. Not unlike the way that the derivative associated to each point of a curve determines the whole curve, K associated to each point characterizes the whole space. In general, K can change from point to point. For the reason to be discussed in Sec.5.4, Luneburg assumed that K is constant for all points in VS under a given condition. In this book, a Riemannian space of constant K is denoted as R. A Euclidean space, denoted as E. is a special case of R. According to $K < 0$, $K = 0$, or $K > 0$, R is hyperbolic space, Euclidean space or elliptic space.

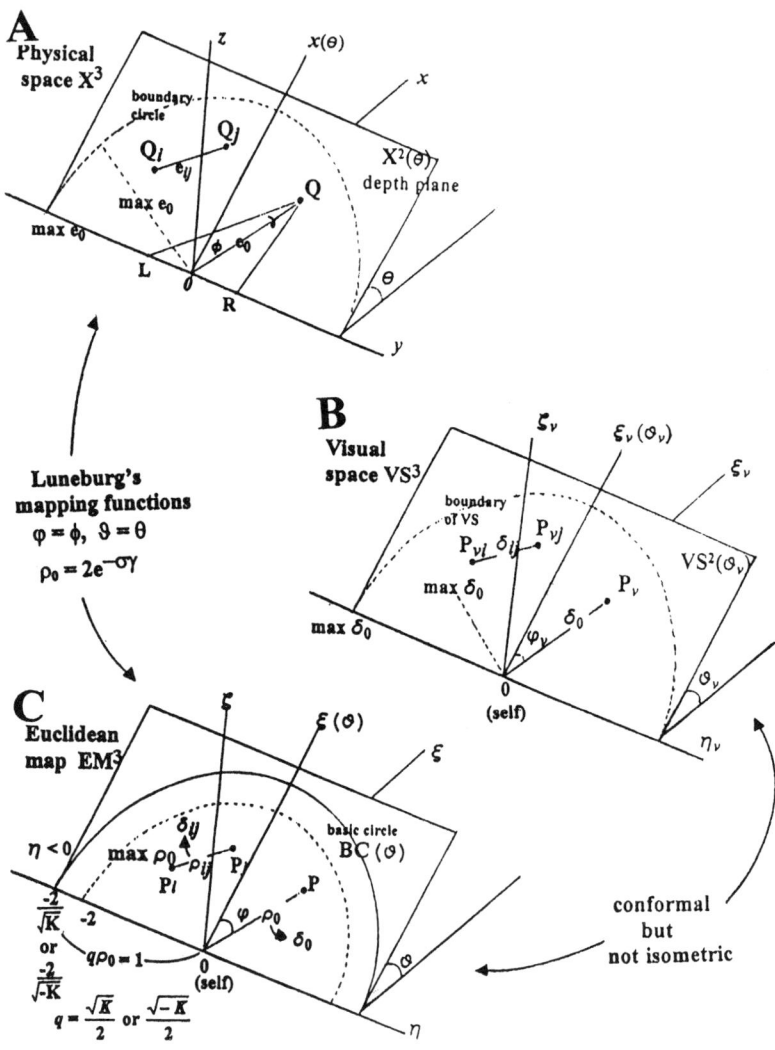

Fig.2.3 Physical space, visual space, and Euclidean map

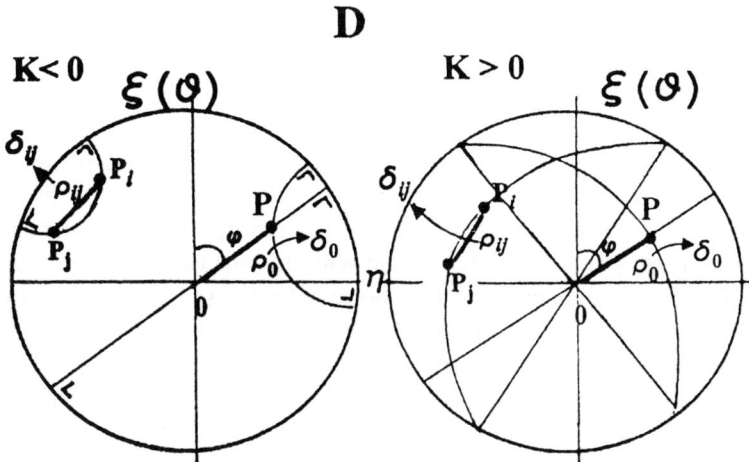

Fig.2.3 (continued) Geodesics in Euclidean map EM

Luneburg regarded that the finding of $\{Q_i\}_D$ being outside $\{Q_i\}_P$ in the horizontal DP implies that $K < 0$ in this VS^2. He was not explicit about whether this assumption covers the entire VS^2 and about whether VS^3 is also hyperbolic. Let us assume that VS^2 under consideration, DP, is an R^2, no matter whether horizontal ($\vartheta_V = 0$) or slanted ($\vartheta_V \neq 0$).

Suppose a person looking at Q in X^3 (Fig.2.3A). Its location is (x, y, z) according to the Cartesian coordinate axes or (γ, ϕ, θ) in terms of the bipolar coordinates. The vertical axis z and the elevation angle θ are added to Fig.1.3I. Fig.2.3B depicts VS of this person. The percept of Q is denoted by P_V and its Cartesian coordinates are (ξ_V, η_V, ζ_V). To describe VS, it is not necessary to use bipolar coordinates, and polar coordinates $(\delta_0, \varphi_V, \vartheta_V)$ are used. The vertical axis ζ_V and ϑ_V are added to Fig.1.3II. The perceived distance of e_{ij} between Q_i and Q_j is denoted as δ_{ij}. In contrast to e_0 and e_{ij} in X, δ_0 and δ_{ij} in VS are latent variables. As will be stated in Sec. 2.2.3, that $\{Q_i\}_D$ lies outside $\{Q_i\}_P$ was demonstrated not only in the horizontal DP ($\theta = 0$) but also in a slanted DP ($\theta > 0$). Any plane having a fixed ϑ_V in Fig.2.3B is assumed to be an R^2. If this $VS^2(\vartheta_V)$ is either hyperbolic ($K < 0$) or elliptic ($K > 0$), some geometrical properties of a perceived figure in $VS^2(\vartheta_V)$ are not quantitatively represented on a sheet of paper, because the sheet is an E^2, 2-D

Euclidean plane, where K = 0. We need Fig.2.3C. All of the three figures in Fig.2.3 will be more fully explained in the next section.

2.2.2. Euclidean Map (EM)

A Riemannian plane R^2 can be represented in E^3. An elliptic plane R^2 (K > 0) is often represented by the surface of a sphere of radius r in E^3. In this case, $K = 1/r^2$ and the distance between points a and b on the sphere along the great circle (the intersection with the surface of a plane passing through a, b, and the center) is the shortest. In that sense, it is the straight line on the sphere R^2, and the arc of the great circle connecting a and b is called the *geodesic*. In the case of a hyperbolic plane (K < 0), most textbooks give as its representation either the surface of a pseudosphere or a saddleback. A geodesic is a curve on this surface. In general, R^m can be embedded in E^M in the similar way, if M is sufficiently larger than m. If $m \leq 3$, R can be represented in E of the same dimensionality in two ways; one is given by Klein by and the other by Poincaré. Let us call the latter, one given by Poincaré, *Euclidian map* of R. It will be abbreviated as EM. Fig.2.3C is EM of VS in Fig.2.3B. The location of a point P is given by (ξ, η, ζ) in terms of the Cartesian coordinates or ($\rho_0, \varphi, \vartheta$) in terms of the polar coordinates. First, general properties of EM are enumerated.

EM1. It is a representation of R, including three cases: K < 0, K = 0, and K > 0, in E of the same dimensionality.

EM2. All angles in R are represented in EM without distortion. This property is called *conformal* representation.

EM3. Unless K = 0, it is impossible to have *isometric* representation of R in EM. A geodesic in R^2 is represented by an arc satisfying a special condition as shown in Fig.2.3D. Denote by δ the length of a geodesic in R and by ρ the Euclidean distance of corresponding chord in EM. Then, δ and ρ are related in simple ways.

$K < 0$	$K > 0$	
$q\delta_{jk} = \sinh^{-1} \dfrac{q\rho_{jk}}{G_j G_k}$	$q\delta_{jk} = \sin^{-1} \dfrac{q\rho_{jk}}{G_j G_k}$	(2.2.1)
$G_j = \sqrt{1 - (q\rho_{0j})^2}$	$G_j = \sqrt{1 + (q\rho_{0j})^2}$	
$G_j = \sqrt{1 - (q\rho_{0k})^2}$	$G_j = \sqrt{1 + (q\rho_{0k})^2}$	
$q = \dfrac{\sqrt{-K}}{2}$	$q = \dfrac{\sqrt{K}}{2}$	(2.2.2)

For radial distances in which $P_j = 0$ and $P_k = P$ and $\rho_{0k} = \rho_0$,

$$q\delta_0 = \tanh^{-1} q\rho_0 \qquad q\delta_0 = \tan^{-1} q\rho_0 \qquad (2.2.3)$$

In Eqs. (2.2.1, 3), sinh and tanh are respectively hyperbolic sine and hyperbolic tangent functions. These are related to ordinary trigonometric sin and tan; $\sin ix = i \sinh x$, $\tan ix = i \tanh x$, where $i = \sqrt{-1}$. If $K = 0$, there is no need to distinguish EM and VS as R and the distance δ_{ij} is equal to or proportional to the Euclidean distance ρ_{ij} between two points P_i and P_j in EM.

To assume that VS is represented as an R means that a perceived distance is magnitude like experience and it behaves in the same way as δ in the above equations does. Unless $K = 0$, a perceptually straight line is represented by an arc. To a creature looking our VS from outside, a geodesic may be a curved line, but for us it is straight and the distance δ_{ij} we see between P_{Vi} and P_{Vj} is related to the chord ρ_{ij} (not accessible to us in VS) as stated above. Perceptual distance δ is a latent variable and we have no natural unit to represent it as a quantitative variable, and the unit of K is not specified yet in Eq. (2.2.2). However, the units of δ and K can be defined together in accordance with Eq. (2.2.1) and Eq. (2.2.3). In order to give an appropriate unit to K, the following properties of $EM^2(\vartheta)$ (Fig.2.3C) must be taken into account.

EM4. When $K \neq 0$, $VS^2(\vartheta_V)$ is represented in EM^2 within a half circle of radius $q\rho_0 = 1$. The circumference in Fig.2.3C is called the *basic circle* and denoted as $BC(\vartheta)$. When $K < 0$, $q\delta_0 = \tanh^{-1} 1 = \infty$ in Eq.(2.2.3), which means that $BC(\vartheta)$ is the set of points at infinity from the self for all directions φ_V. VS is closed and all δ_0 are finite (VS3 in Sec.1.1.1). Hence the boundary of $VS^2(\vartheta_V)$ must be inside of $BC(\vartheta)$. When $K > 0$, as will be discussed later, $BC(\vartheta)$ is the set of points at which all arcs representing perceptually straight lines in a given direction intersect with each other. In other words, there cannot be "parallel" lines if we take the entire $BC(\vartheta)$. Hence, $VS^2(\vartheta_V)$ in which the S perceives P-alley must be represented inside of $BC(\vartheta)$. Namely, in either case, $K < 0$ or $K > 0$, max ρ_0, corresponding to max δ_0, must be represented within $BC(\vartheta)$ of the radius $q\rho_0 = 1$ or $\rho_0 = 1/q = 2/\sqrt{\pm K}$ It is convenient to define the scale of ρ in EM^2 such that max $\rho_0 = 2$, because the possible range of K is limited as follows from Eq.(2.2.2).

$$-1 < K < 0 \text{ when } K < 0, \qquad 0 < K < 1 \text{ when } K > 0. \qquad (2.2.4)$$

The dotted half circle in Fig.2.3C represents max $\rho_0 = 2$. This is fixed by the definition. It depends upon K how far the dotted circle and $BC(\vartheta)$ are apart. When $K = 0$, max ρ_0 is still 2 but $BC(\vartheta)$ goes to infinity.

EM5. The left figure in Fig.2.3D is a schematic picture of $EM^2(\vartheta)$ when $K < 0$. All perceptual straight lines are represented by geodesic circles that are perpendicular to $BC(\vartheta)$ at its both ends. The perceived straight line passing through P_{Vi} and P_{Vj} in Fig.2.3B is represented by the small geodesic circle through P_i and P_j and δ_{ij} is given from the length of chord ρ_{ij} (Eq.2.2.1). As shown by an example passing P, both ends of a geodesic circle do not need to meet $BC(\vartheta)$ on the side where $\xi(\vartheta) > 0$. The radial straight line from O to P is orthogonal to $BC(\vartheta)$ at its both ends. Hence, this line is the geodesic and its length ρ_0 determines δ_0 (Eq. 2.2.3).

The right figure in Fig.2.3D is a schematic picture of $EM^2(\vartheta)$ when $K > 0$. All perceptual straight lines are represented by geodesic circles intersecting $BC(\vartheta)$ at antipodal points. That means that two intersecting points are connected by a diameter.

All radial lines extending from the origin represent perceived radial straight lines from the self in $VC^2(\vartheta_V)$ both when $K < 0$ and $K > 0$, because these are orthogonal to $BC(\vartheta)$ when $K < 0$ and antipodal when $K > 0$. Mathematical explanation of EM and derivations of all equations will be given in Sec. 2.4.2.

The two curves of P-alley passing through $P_1(\xi_1(\vartheta), \pm\eta_1)$ are perceptually parallel, which means that each is parallel with $\xi(\vartheta)$. From Fig.2.3D, it is easy to see that when $K < 0$ there are many geodesics that pass through P_1 and do not intersect $\xi(\vartheta)$. Hence, the next problem is how to define P-curves.

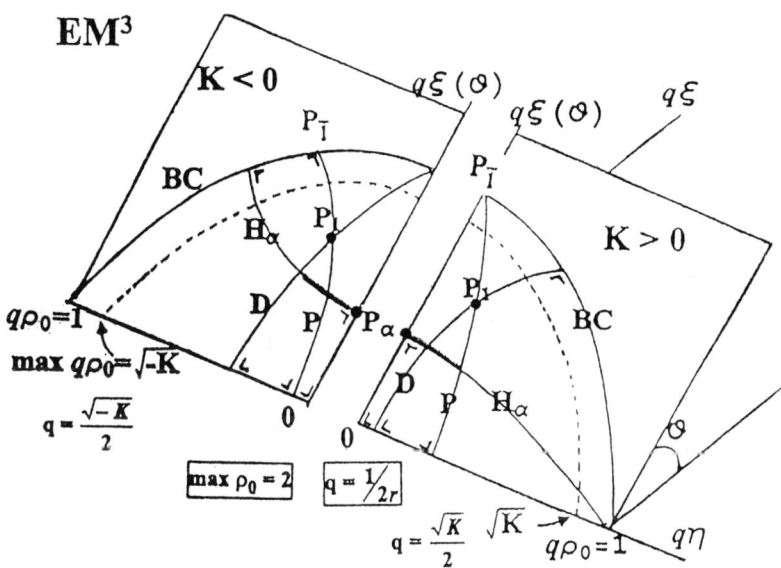

Fig.2.4 P-and D-alleys, H-curves in $EM^2(\vartheta)$

2.2.3. *Equations of P-and D-alleys, H-curves in EM^2*

At first, Luneburg (1947) developed theoretical equations for P-and D-alleys directly in the physical space X^2. In the second paper of 1950, published after his untimely death in 1949, all equations were first developed in EM^2 and then the curves were projected to X^2. This is an easier way to make explicit the ideas behind equations. Hence, equations of P-and D-alleys and H-curves are first given in $EM^2(\theta)$. These equations hold both when $K < 0$ and $K > 0$, and the curves defined by these are shown in Fig.2.4. In order to depict these curves, it is more convenient to use coordinates $(q\xi, q\eta)$ and $q\rho$. Since each theoretical curve is symmetric with regard to the $\xi(\vartheta)$ axis, the curve in one side only is shown. In Fig.2.4, the left side shows the left side curves when $K < 0$ and the right side shows the right side curves when $K > 0$. Each of these equations defines a geodesic circle in $EM^2(\vartheta)$.

P-alley

$$K<0 \quad (q\rho_0)^2 - C_p(q\eta) +1 = 0, \quad C_p = \frac{1+(q\rho_{01})^2}{q\eta_1} \qquad (2.2.5)$$

$$K>0 \quad (q\rho_0)^2 + C_p(q\eta) -1 = 0, \quad C_p = \frac{1-(q\rho_{01})^2}{q\eta_1} \qquad (2.2.6)$$

D-alley

$$K<0 \quad (q\rho_0)^2 + C_D(q\eta) - 1 = 0, \quad C_D = \frac{1-(q\rho_{01})^2}{q\eta_1} \qquad (2.2.7)$$

$$K>0 \quad (q\rho_0)^2 - C_D(q\eta) + 1 = 0, \quad C_D = \frac{1+(q\rho_{01})^2}{q\eta_1} \qquad (2.2.8)$$

H_α-curve

$$K<0 \quad (q\rho_0)^2 - C_{H\alpha}(q\xi(\vartheta)) +1 = 0, \quad C_{H\alpha} = \frac{1+(q\xi_\alpha(\vartheta))^2}{q\xi_\alpha(\vartheta)}$$

$$\qquad (2.2.9)$$

$$K>0 \quad (q\rho_0)^2 + C_{H\alpha}(q\xi(\vartheta)) -1 = 0, \quad C_{H\alpha} = \frac{1-(q\xi_\alpha(\vartheta))^2}{q\xi_\alpha(\vartheta)}$$

$$\qquad (2.2.10)$$

Each side member of P-alley passing through P_1 is perceptually straight and parallel with $\xi_V(\vartheta_V)$-axis in VS. When $K < 0$, straight lines in $VS^2(\vartheta_V)$ are represented in EM by arcs orthogonal to $BC(\vartheta)$ at its both ends (EM5). There are many geodesic circles passing through P_1 that are orthogonal to $BS(\vartheta)$, one end on the side $\xi(\vartheta) > 0$ and the other end on the side $\xi(\vartheta) < 0$, and do not intersect with $\xi(\vartheta)$-axis. Hence, Luneburg introduced a beautiful definition of being parallel in $VS^2(\vartheta_V)$. In the left side of Fig.2.4, the arc P is such a geodesic circle that is orthogonal to the qη-axis. The arc P shares this property with $\xi(\vartheta)$. Because VS and EM are conformal (EM2), both being orthogonal to the qη-axis implies that these two perceptual lines stretch together in parallel straight ahead from the η_V-axis to the boundary of VS.

When $K > 0$, the right side member of P-alley passing through P_1 is represented in EM by the geodesic circle P. This curve P and $\xi(\vartheta)$ both meet $BC(\vartheta)$ at their antipodal points and are orthogonal to the qη-axis. In $VS^2(\vartheta_V)$, the two straight lines stretch from the η_V-axis straight ahead and meet only at one point beyond the boundary of $VS^2(\vartheta_V)$.

An H_α-curve in $EM^2(\vartheta)$ represents the frontoparallel straight line passing through $P_{V\alpha}$ in $VS^2(\vartheta)$. Being frontoparallel means that the line is parallel to the qη_V-axis. Namely, in $EM^2(\vartheta)$ the relation between H_α-curve and the qη-axis must be the same as the relation between P-curve and the q$\xi(\vartheta)$-axis. In Fig. 2.4, the arc H_α passing through P_α satisfies this condition. The arc H_α on the left is orthogonal to $BC(\vartheta)$ and to the q$\xi(\vartheta)$ axis, whereas the arc H_α on the right meets the qη-axis on $BC(\vartheta)$. Hence, Eqs.(2.2.9) and (2.2.10) are equivalent to Eqs.(2.2.5) and (2.2.6) if $\xi(\vartheta)$ is replaced by η.

A D-alley in $VS^2(\vartheta_V)$ means the loci of two points having a constant separation distance in the direction of η_V-axis. Namely points on a member of D-alley must have a constant distance from the $\xi_V(\vartheta)$ axis. In $EM^2(\vartheta)$ the distances must be defined along H_α-arcs. In Fig.2.4, D-curves satisfy this condition. When P_α moves on the $\xi(\vartheta)$-axis, the distance δ from the corresonding P on the heavy arc remains constant. Though D-curves are arcs in $EM^2(\vartheta)$, these do not represent stright lines in $VS^2(\vartheta_V)$. The arc D in the left ($K < 0$) is not orthogonal to $BC(\vartheta)$ and the arc D on the right ($K > 0$) does not meet $BC(\vartheta)$ at its antipodal points. When a D-alley was constructed and the whole set $\{Q_i\}_D$ was

32 *Global Structure of Visual Space*

presented to the S, sometimes the S spontaneously commented that the series of P_V in each side of the $\xi_v(9)$ did not appear straight.

When $K < 0$, the arc D lies outside the arc P whereas, when $K > 0$, the relationship is reversed. It is interesting to see that the arc P when $K < 0$, and the arc D when $K > 0$ are the same, corresponding to that Eqs. (2.2.5) and (2.2.8) are the same. Similarly, the arc D when $K < 0$ and the arc P when $K > 0$ are the same, corresponding to that Eqs. (2.2.6) and (2.2.7) are the same. Derivations of all equations are given in Sec. 2.4.2.

Each equation, (2.2.5) to (2.2.10), defines a circle having the following center and radius.

P-alley   center $O(P_1)$ on $q\eta$-axis   radius

$K < 0$ $\quad q\dot\eta_P = \dfrac{C_P}{2} \qquad\qquad r_P = \sqrt{(q\dot\eta_P)^2 - 1}$ (2.2.5')

$K > 0$ $\quad q\dot\eta_P = \dfrac{-C_P}{2} \qquad\quad r_P = \sqrt{(q\dot\eta_P)^2 + 1}$ (2.2.6')

D-alley   center $O(P_1)$ on $q\eta$-axis   radius

$K < 0$ $\quad q\dot\eta_D = \dfrac{-C_D}{2} \qquad\quad r_D = \sqrt{(q\dot\eta_D)^2 + 1}$ (2.2.7')

$K > 0$ $\quad q\dot\eta_D = \dfrac{C_D}{2} \qquad\qquad r_D = \sqrt{(q\dot\eta_D)^2 - 1}$ (2.2.8')

H_α-curve   center on $q\xi$-axis   radius

$K < 0$ $\quad q\dot\xi_{H\alpha} = \dfrac{C_{H\alpha}}{2} \qquad\quad r_{H\alpha} = \sqrt{(q\dot\xi_{H\alpha})^2 - 1}$ (2.2.9')

$K > 0$ $\quad q\dot\xi_{H\alpha} = \dfrac{-C_{H\alpha}}{2} \qquad r_{H\alpha} = \sqrt{(q\dot\xi_{H\alpha})^2 + 1}$ (2.2.10')

The center is defined positive when it is on the same side with the fixed point P_1. Fig.2.13 to be used in Sec. 2.4.2 shows the centers $O(P_1)$ and radii of P-alleys when $K < 0$ and $K > 0$. In Figs.2.4, 2.13 and other figures, the point at which P-alley intersects BC is denoted as $P_{\bar{1}}$. It will be clear that P-alley curve is orthogonal with BC at $P_{\bar{1}}$ when $K < 0$ and $P_{\bar{1}}$ is an antipodal point when $K > 0$.

All experimental results $\{Q_i\}_P$, $\{Q_i\}_D$, and $\{Q_i\}_H$ are obtained in $X^2(\theta)$, and all the curves discussed above are defined in $EM^2(\vartheta)$. Hence, in order to compare theoretical curves with experimental results, we have to map all the curves in $EM^2(\vartheta)$ back to $X^2(\theta)$.

2.3. Theoretical Curves in X^2

2.3.1. Luneburg's Mapping Functions

Luneburg proposed the following relations between $Q(\gamma, \phi, \theta)$ in X^3 and $P(\rho_0, \varphi, \vartheta)$ in EM^3 (Fig.2.3). These will be called *Luneburg's mapping functions* (throughout this book, ϕ is defined in a slightly different way from his definition, see Fig.5.9).

$$\rho_0 = g(\gamma) = 2e^{-\sigma\gamma} \tag{2.3.1}$$

$$\varphi = \phi, \quad \vartheta = \theta \tag{2.3.2}$$

In X^3, it is easier to specify the position of a stimulus as $Q(x, y, z)$ according to the Cartesian coordinates. The bipolar coordinates are obtained by

$$X = 2x/f, \quad Y = 2y/f, \quad Z = 2z/f$$

$$\gamma = \tan^{-1} \frac{2\sqrt{X^2 + Z^2}}{X^2 + Y^2 + Z^2 - 1} \tag{2.3.3a}$$

34 Global Structure of Visual Space

$$\phi = \frac{1}{2}\tan^{-1}\frac{2Y\sqrt{X^2+Z^2}}{X^2+Z^2-Y^2+1} \qquad (2.3.3b)$$

$$\theta = \tan^{-1}\frac{Z}{X} \qquad (2.3.3c)$$

where f is the pupil distance of the S. It is arbitrary in what unit to measure physical varables, x, y, z, because these are normalized by another physical length f/2.

Once a theoretical point $P(\rho_0, \varphi, \vartheta)$ is obtained in EM^3, it will be easier to use the Cartesian coordinates to plot P in EM^3.

$$\xi(\vartheta) = \rho_0 \cos\varphi \qquad (2.3.4a)$$

$$\xi = \xi(\vartheta) \cos\vartheta \qquad (2.3.4b)$$

$$\eta = \rho_0 \sin\varphi \qquad (2.3.4c)$$

$$\zeta = \xi(\vartheta) \sin\vartheta \qquad (2.3.4d)$$

If σ is known, $P(\rho_0, \varphi, \vartheta)$ in EM can be mapped in X as $Q(x, y, z,)$.

$$\gamma = \frac{1}{\sigma}\ln\frac{2}{\rho_0} \quad \text{(from Eq.2.3.1)} \qquad (2.3.5a)$$

$$x = \frac{f}{2}\frac{\cos 2\phi + \cos\gamma}{\sin\gamma}\cos\theta \qquad (2.3.5b)$$

$$y = \frac{f}{2}\frac{\sin 2\phi}{\sin\gamma} \qquad (2.3.5c)$$

$$z = \frac{f}{2}\frac{\cos 2\phi + \cos\gamma}{\sin\gamma}\sin\theta \qquad (2.3.5d)$$

Eqs. (2.3.2) mean that directional angles ϕ and θ from O in X^3 are preserved as φ and ϑ in EM^3. Since VS^3 and EM^3 are conformal (EM2); $\varphi_V = \varphi$ and $\vartheta_V = \vartheta$. Hence, Eqs. (2.3.2) make VS^3 and X^3 to be conformal insofar as these directional angles are concerned.

$$\varphi_V = \varphi = \phi, \qquad \vartheta_V = \vartheta = \theta \qquad (2.3.6)$$

Namely the assumption (2.3.2) is consistent with the veridicality of VS in the neighborhood of the self (VS4 in Sec.1.1.1).

The mapping functions (2.3.1) and (2.3.2) presuppose that P_V is localized with regard to the self in VS (*ego-centric* localization). For P's under the influence of framework other than the self, (2.3.1) and (2.3.2) do not hold. For instance, in Fig.1.2B, the position of the point $Q(\gamma, \phi, \theta)$ remains the same during P appears to move by induction of the movement of rectangle. That P_V moves to the left in VS logically means that its φ_V is changed with regard to the self. However, this P is localized with regard to the perceived rectangle but not to the self, and hence its ϕ is not functional in this case. If the rectangle appears to move to the right in Fig.1.2, it means that the rectangle is localized with regard to the self and the change of its ϕ causes the change of φ_V.

Eq.(2.3.1) is the assumption on how ρ_0, the independent variable in Eq.(2.2.3) in Sec. 2.2.2, changes when Q moves away from the body in X^3. When the distance e_0 from body to Q becomes larger, γ gets smaller. The decrease of γ is related to e_0 and the pupil distance f (2.3.3). From Eq.(2.3.1),

$$\frac{d\rho_0}{d\gamma} = -\sigma \rho_0$$

and ρ_0 determines the perceptual radial distance δ_0. From (2.2.3). both for $K < 0$ and $K > 0$,

$$\frac{d\delta_0}{d\rho_0} = \frac{1}{1+(q\rho_0)^2} > 0 \qquad (2.3.7)$$

and hence,

$$\frac{d\delta_0}{d\gamma} = \frac{d\delta_0}{d\rho_0}\frac{d\rho_0}{d\gamma} = \frac{-\sigma\rho_0}{1+(q\rho_0)^2} \tag{2.3.8}$$

The parameter σ in Eq.(2.3.1) means the sensitivity of how the perceptual radial distance δ_0 increases when e_0 increases and γ decreases. The range in which γ is effective in determining δ_0 is quite limited (Sec.1.2.1). In fact, Eq. (2.3.7) becomes practically zero if $e_0 > 15$ m, irrespective of the pupil distance f, the sign of K and the value of σ. The Luneburg's mapping function (2.3.1) assumes that, in determining δ_0, the physical distance to Q from O, e_0, is effective only through the convergence angle γ and Eq. (2.3.1) is meaningful only in the neighborhood of the body. When the ground or other framework is visible, however, Q is not egocentrically localized, and e_0 can be the direct physical counterpart of δ_0 (Sec.4.2).

In the Luneburg model, K and σ are only two parameters. The former determines the structure of EM and the latter is involved in the mapping function (2.3.1). The assumed form (2.3.1) has one more important characteristic of being consistent with so-called iseiconic transformation (Hardy, et al., 1953; Blank, 1957; Eschenburg, 1980). This problem will be discussed in Sec.5.2.3. The characteristics of Luneburg's mapping functions are summerized as follows.

MF1. Eqs.(2.3.1 and 2) are egocentric and meaningful only when VS is frameless.

MF2. The three variables, γ, ϕ, θ, are assumed to have their effects independently. Namely, the same form $g(\gamma)$ in (2.3.1) is assumed to hold for any value of ϕ and θ. Putting together this assumption and the assumption of K being constant, we can say that VS is assumed to be homogeneous and isotropic (independent of direction), at least in the neighborhood of the self.

MF3. The forms of (2.3.1) and (2.3.2) are *a priori* assumptions. As mentioned in Sec.2.2.1, the assumption of K being constant is made on the basis of a set of rationales. Eqs. (2.3.2) are necessary for the veridicality of VS. As to Eq. (2.3.1), however, there are no such rationales.

All experiments on P-and D-allyes and H-curves so far referred to (*e.g.*, Fig.2.1) were performed in X^2 in which the Luneburg's mapping functions are expected to hold. The procedure to determine the curves by estimating the optimum values of K and σ is explained in the next section.

2.3.2. Equations of P-and D-alleys and H-curves in X^2

The same equation, (2.2.5) and (2.2.7), holds for P-alley when $K < 0$ and D-alley when $K > 0$. This equation can be written as

$$[1 + (q\rho_0)^2] - [1 + (q\rho_{01})^2]\frac{q\eta}{q\eta_1} = 0$$

Using $\eta = \rho_0 \sin\varphi$ (Eq.2.3.4c), $\varphi = \phi$ (Eq.2.3.6), and multiplying both sides by $1/2q\rho_0$, we have

$$\frac{1+(q\rho_0)^2}{2q\rho_0} = \frac{1+(q\rho_{01})^2}{2q\rho_{01}} \frac{\sin\phi}{\sin\phi_1} \qquad (2.3.9)$$

Let us consider the following functions

$$e^{-(\sigma\gamma+k)} = e^{-\sigma\gamma} e^{-k}, \quad k = \ln(1/2q), \text{ and } \rho_0 = 2e^{-\sigma\gamma} \quad (\text{Eq.2.3.1})$$

then, $e^{-k} = 2q$ and $e^{-\sigma\gamma} = \frac{1}{2}\rho_0$, and hence

$$q\rho_0 = e^{-(\sigma\gamma+k)}$$

The left side of (2.3.9) can be written as

$$\frac{1+e^{-2(\sigma\gamma+k)}}{2e^{-(\sigma\gamma+k)}}$$

This is the form that can be written as cosh $(\sigma\gamma + k)$ because hyperbolic cosine is

$$\cosh x = \frac{e^x + e^{-x}}{2} = \frac{1 + e^{-2x}}{2e^{-x}}$$

Hence, (2.3.9) is written in terms of cosh $(\sigma\gamma + k)$. Similarly, the same equation, (2.2.6) and (2.2.8), for D-alley when K < 0 and P-alley when K < 0 can be written in terms of sinh $(\sigma\gamma + k)$. Hyperbolic sine is

$$\sinh x = \frac{e^x - e^{-x}}{2} = \frac{1 - e^{-2x}}{2e^{-x}}$$

P-alley (K<0) and D-alley (K>0),

$$\frac{\cosh(\sigma\gamma + k)}{\cosh(\sigma\gamma_1 + k)} = \frac{\sin\phi}{\sin\phi_1} \qquad (2.3.10)$$

D-alley (K< 0) and P-alley (K > 0)

$$\frac{\sinh(\sigma\gamma + k)}{\sinh(\sigma\gamma_1 + k)} = \frac{\sin\phi}{\sin\phi_1} \qquad (2.3.11)$$

H_α-curve (K < 0 and K > 0)

K < 0   K > 0

$$\frac{\cosh(\sigma\gamma + k)}{\cosh(\sigma\gamma_\alpha + k)} = \cos\phi \qquad \frac{\sinh(\sigma\gamma + k)}{\sinh(\sigma\gamma_\alpha + k)} = \cos\phi \qquad (2.3.12)$$

$$k = \ln\frac{1}{\sqrt{-K}} \qquad k = \ln\frac{1}{\sqrt{K}} \qquad (2.3.13)$$

The fixed point Q_α is on the ξ axis. Hence, $\phi_\alpha = 0$, $\cos \phi_\alpha = 1$ and $\xi_\alpha = \rho_{0\alpha}$.

For a set of data points Q_i (γ_i in radian, ϕ_i) in $X^2(\theta)$, parameters in the above equations are σ and k. From k, we can have K (Eq.2.3.13). Once we know the values of σ and k, for any value of γ and hence ρ_0 (Eq.2.3.1), we can have the value of $\phi(\gamma)$ through these equations. For ($\rho_0(\gamma)$, $\phi(\gamma)$, θ), the corresponding $Q(x, y, z)$ can be determined through (Eqs.2.3.5). Three sets of curves, P-and D-alleys and H_α-curves in Fig.2.1 were obtained in this way where $\vartheta = \theta = 0$ and $\cos \theta = 1$. The values of K and σ used for this purpose are shown in Fig.2.1.

Eqs. (2.3.10 - 12) are expressed in the form that the right sides do not have unknown parameters. Nowadays it may be easy to obtain the optimum values of σ and k through a nonlinear regression program in any software package. When the curves in Fig.2.1 were fitted, the situation was different. Luneburg proposed two procedures to estimate K and σ. One is to perform two additional experiments and the other is to use the tangent lines of the linear parts of data in the vicinity of Q_1. These two procedures were tested in Indow et al. (1962a). The results of the first procedure based upon the so-called 3-and 4-point experiments were not satisfactory. In the second procedure, the tangent lines must be visually determined and the estimation is unstable. The computations in these procedures could be carried out by a desk calculator. The optimization of values of σ and k by the method of steepest descent is an iterative procedure and needs a computer. In the same article, we tried this procedure and it gave the best results. All values of K and σ reported in this article, such as ones in Fig. 2.1, were obtained by this procedure. The most recent version of that program is explained in Appendix 2 in Indow and Watanabe (1984a).

2.3.3. *Comments on Results of Alley Experiments*

Frontoparallel settings $\{Q_i\}_{H\alpha}$ are fitted by concave curves when Q_α is close to the S and by convex curves when Q_α is far (Fig.2.1). This fact is well known (Ogle, 1964). In EM^2, for all fixed points P_α, H_α-curves (Eq.2.3.12) are convex when $K < 0$ and concave when $K > 0$ (Fig.2.4). However, when mapped to X^2 through the Leneburg's mapping functions, these are of the form as shown in Fig.2.1, no matter whether $K < 0$

or K > 0. The sign of K only affects on the position of inflection point at which concavity changes to convexity. Hence, the frontoparallel settings $\{Q_i\}_{H\alpha}$ are not informative to differentiate whether K is positive or negative.

All experimental results in my laboratory showed that D-alley $\{Q_i\}_D$ lies outside P-alley $\{Q_i\}_P$ and K turned out to be negative. The values of K and σ obtained before 1979 (more than 50 cases) are listed in Table 1 in Indow and Watanabe (1984a). The situation is the same in Table II (30 cases) in Zajaczkowska (1956a). In other experimental studies, some cases were reported that do not match the assumption of K < 0. A most noticiable case is the experiment carried out in the large outdoor fields (Battro, et al., 1976). They used wooden stakes colored yellow as Q's that were moved by the experimenter according to the direction of the S. P- and D-alleys were constructed with each of $Q_1(x_1, y_1)$ that was varied in 9 ways. The largest case was $Q_1(240 m, 24 m)$. The total number of pairs, $\{Q_i\}_P$ and $\{Q_i\}_D$, was 203. Equations having K and σ were fitted and they did not report values of σ. As to K, 52 pairs were with K < 0 and 38 pairs were with K > 0. The remaining 113 pairs were classified as "nonregular that were not suited for computation". It is a problem to use the Luneburg's mapping functions. That this is not a frameless VS is not the problem, but the size of $\{Q_i\}$ is too large to use Eq. (2.3.1). In addition to this problem, I express some doubt as to whether all S's really understood the task because of the large number of "nonregular" cases. A total of 46 S's participated. The authors did not describe about S's, except that they are "male and female adults". It is not easy to make it clear to this large number of S's what is really meant by "perceptually parallel" or by "perceptually equi-distant".

Hardy et al. (1951) reported, in addition to 9 S's whose results were fitted by equations of K < 0, results of two anomalous groups of S's. One group (2 S's) constructed both $\{Q_i\}_P$ and $\{Q_i\}_D$ parallel to the x-axis. Fig.2.5A shows one case replotted from Chart2 of their article. In a discussion with these S's, the authors concluded that they "had been unable to dissociate physical and sensory parallelism". After the discussion, one S constructed P- and D-alleys that were similar to Fig.2.1. Two S's in the other group "had had training in art and the portrayal of vanishing point and central projection perspective". Both constructed $\{Q_i\}_P$ and $\{Q_i\}_D$ in which the two series of Q_i were both on the straight line connecting Q_1 with R or L. Fig.2.5 B shows one case replotted from Chart3 of their article. When we performed alley experiments, we instructed S's about the difference between "perceptually parallel or equidistant" and

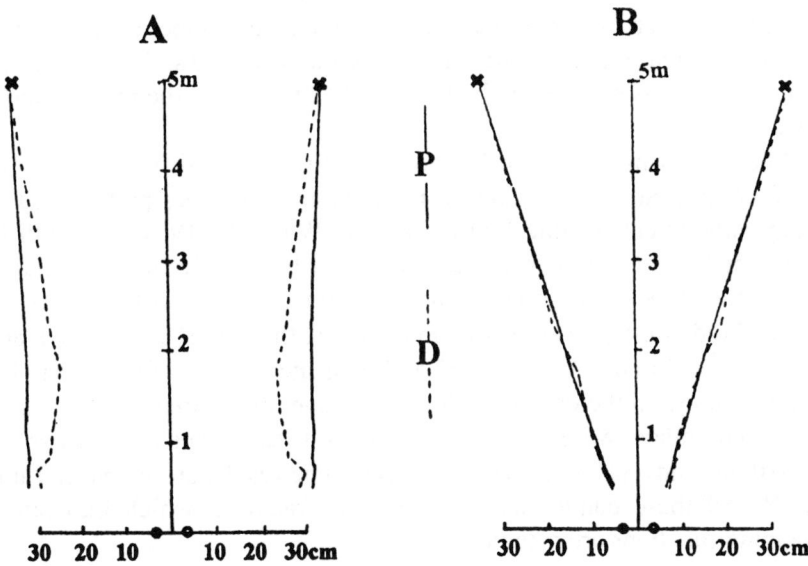

Fig.2.5 Two anomalous examples from the experiment of Hardy *et al.* (1951) The scale of y-axis is enlarged

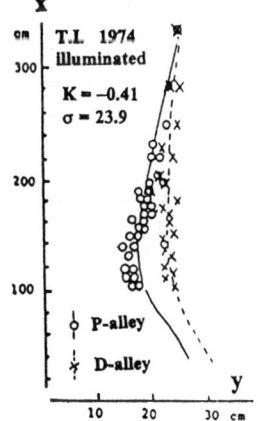

Fig.2.6 An example of alleys with a large number of Q's. Replotted from Fig.5 in Indow (1982). The scale of y-axis is enlarged

"physically parallel or equidistant" in detail. However, it was carefully avoided to suggest anything about the possible discrepancy between $\{Q_i\}_P$ and $\{Q_i\}_D$. If the S's understand the task, they construct such $\{Q_i\}_P$ and $\{Q_i\}_D$ that are well fitted by Eqs.(2.3.10) and (2.3.11) of $K < 0$. Fig.2.6 shows an example of alleys (the right side member only) that I as a S constructed with extraordinarily large number of Q's. Notice that the scale of y-axis is enlarged in both Figs.2.5, 2.6 to show the patterns more clearly.

Alley experiments in the horizontal DP($\theta = 0$) with stationary $\{Q_i\}$ were conducted under various conditions and some S's participated in many experiments (Table 1 in Indow and Watanabe 1984a). Stimuli Q_i were small light points in the dark or small black objects in the illuminated frameless space. In laboratory experiments, the coordinates (x_1, y_1) of the fixed point Q_1 were varied in the range that $150 \leq x_1 \leq 417$ cm and $12 \leq |y_1| \leq 76$ cm. Values of K and σ of the same S's for comparable sizes of alleys in the dark and illuminated conditions are shown in Table 2.1A. The alleys were also constructed in a dark gymnasium where x_1 =1610 cm and $|y_1| = 150$ cm or 75 cm. The results are given in Table 2.1.B. All these conditions are within the range in which Luneburg's mapping functions are meaningful.

Table 2.1 Examples of K and σ. **A**: in the laboratory.
B: in the dark gymnasium.

A	Dark		Illuminated	
	K	σ	K	σ
T.I.	-0.30	11.0	-0.38	19.8
T.I	-0.24	12.0	-0.41	23.9
K.M.	-0.40	15.0	-0.24	23.6
L.S.	-0.29	8.9	-0.27	24.6
T.F.	-0.29	12.9	-0.37	21.9
S.A.	-0.41	17.5	-0.14	24.7
N.Y.	-0.12	9.7	-0.30	18.6
S.T.	-0.34	10.1	-0.24	13.7

B In the dark gymnasium

	$x_0 \quad \pm y_0$ cm	K	σ
T.I.	1610, 150	−0.68	45.1
	1610, 75	−0.80	31.2
E.I.	1610, 150	−0.72	30.9
	1650, 75	−0.74	29.2
T.K.	1650, 150	−0.86	28.6
	1650, 75	−0.81	31.8

Findings in Table 2.1 are summarized as follows.

Alley 1. In laboratory experiments Table 2.2A, most individual values of K were in the range between −0.12 and −0.41 for all sizes of alleys in the dark and illuminated spaces. It was expected that illuminated VS might be closer to Euclidean structure and −K might be closer to 0. It was not the case, however. What changed between the two conditions was σ. Systematically it is larger in the illuminated condition, which implies that the same small difference in the convergence angle γ causes a larger change of perceptual radial distance $δ_0$ than in the dark (Eq. 2.3.8).

Alley 2. Values of −K and σ of alleys in the dark gymnasium are listed in Table 2.2B. In this case, it was impossible to have illuminated frameless VS. Though the experiment was performed at night and all lights were turned off, it was inevitable that, due to the dark adaptation, the vague image of the surround became visible if the S stayed long in the dark. Whenever an alley was constructed, the S was exposed to light in order to recondition the eyes, during which the S was prevented from seeing the alley. Clearly −K was further far from 0 and σ was larger compared with those in dark or illuminated laboratory room. As a matter of course, in the gymnasium the S perceived more extended alleys. Hence, it was tried to construct alleys in the dark room in which a background is visible behind the furthest pair P_{V1}. The results of these experiments are discussed in the next section.

Alleys were constructed in non-horizontal DP and the results were essentially the same (Indow, *et al.*, 1962b; Table 1 in Indow and Watanabe 1984a).

2.3.4. Comments on Values of $-K$ and σ

Euclidean map EM has the only one parameter K (or q). Unless we are concerned with how $Q(\gamma, \phi, \theta)$ in X is represented as $P(\rho_0, \varphi, \vartheta)$ in EM, σ does not play any role in EM. One role of $-K$ is to tell how far the two curves, P_P and P_D, appear to be apart in VS. Once $P_1(q\rho_{01}, \varphi_1, \vartheta_1)$ is fixed in EM, the alley curves are determined by Eqs.(2.2.5) and (2.2.7) in which σ is not involved. Fig.2.7 shows P- and D-curves passing through $P_1(q\rho_{01}, \varphi_1, \vartheta_1)$ in EM ($q\eta > 0$) when $K = -0.30$, where $q\rho_{01} = 0.428$, $\varphi_1 = 0.100$, and ϑ_1 can be any value. The perceptual discrepancy between two δ's between two points, P_{VL} and P_{VR}, one in P- alley and one in D-alley, can be represented by the following index r_α. Six points P_α are taken on the $q\xi$-axis with the intervals of $q\xi_1/6$, and the intersecting points of H_α-curves (Eq.2.2.10) with P-and D-alleys are denoted as P_P and P_D. Using Eq.(2.2.1), we can have the ratio

$$r_\alpha = \frac{\delta(P_{VL}, P_{VR})_P}{\delta(P_{VL,} P_{VR})_D}$$

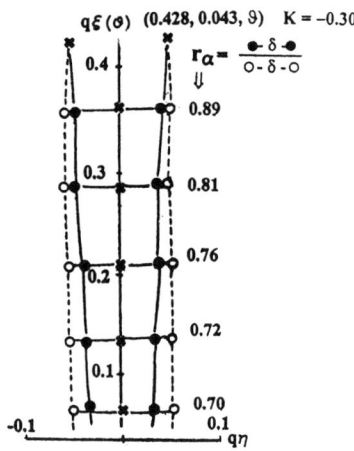

Fig.2.7 How P-and D-alleys are separated in VS is determined by K

These values are given in Fig.2.7. When S's were asked to assess the ratio of these two δ's with $\{Q_i\}_P$ and $\{Q_i\}_D$ constructed by themselves, they gave ratios close to r_α that was based on their values of K and σ. The latter is necessary only to define ρ_{01} from $\gamma(e_{01})$ (Eq.2.3.1). The discrepancy between P-and D-alleys in VS is real in this sense also. Experiments to ask ratios of δ's in more extended ways were performed (Sec. 3.2.1). The other role of $-K$ is to tell how close the boundary max $q\rho_0$ of VS is to the basic circle ($q\rho_0 = 1$). When $K < 0$, as K becomes closer to -1, VS is represented in a larger part of EM. If K is close to 0, VS is represented in a small half circle around 0 and non-Euclidianity of EM becomes meaningless. If $K = -0.30$ and -0.80, max $q\rho_0 = 0.548$ and 0.894 respectively. Notice that max $q\rho_0 < 1$ (Fig.2.4).

In frameless conditions, it is not clear for the S how far VS is extended beyond the configuration of percepts $\{P_{Vi}\}$. It is possible, however, to relate max δ_0 to the radial distance δ_{01} to a point P_{V1}, if its location $q\rho_{01}$ in EM is known. From Eq. (2.2.3), when $K < 0$,

$$R = \frac{\max q\delta_0}{q\delta_{01}} = \frac{\tanh^{-1}\sqrt{-K}}{\tanh^{-1}\left(0.5\sqrt{-K}\rho_{01}\right)} \tag{2.3.14}$$

Table 2.2 summarizes values of R for P_1 of alleys under various conditions. By the use of the values of K and σ obtained in alley experiment, $q\rho_{01}$ and R were determined for each S under the respective conditions. Geometric means of R of 2 or 3 S's, $\overline{R}$, are given in the last column. Conditions 1 to 3 were already described. In Conditions, 4.2, 4.3, and 6.1-6.3, the dimly illuminated checker-board background was placed behind Q_1 at two different distances. In front of this background, P-and D-alleys were constructed with stationary Q's (Condition 4) and with two moving Q's starting from the position of Q_1 (Condition 6). In conditions 5.2 - 5.4, one of the three moving patterns was continuously projected on the screen at 425 cm and the S constructed P-and D-alleys in front of this background. More detailed results were given in Fig.3 of Indow (1997). A few comments on the value of $\overline{R}$ will suffice herein.

In Tables 2.1A and B, K and σ were different between Conditions 1 and 3, but $\overline{R}$ was not too different between these two dark conditions. On the other hand, $\overline{R}$ was clearly larger in Condition 2 than in Condition 1. Between these Conditions, K did not systematically change but σ

Table 2.2 Values of R (2.3.14) under various conditions

Condition		x_1 cm	$\pm y_1$ cm	S's	$\overline{R}$
1.	Light points in dark	417	40	3	1.28
2.	Black balls under illumination	355	30	3	1.70
3.1	Light points in dark gym.	1610	150	3	1.38
3.2		1610	75	3	1.36
4.1	Light points in dark	409	40	2	1.52
4.2	background at 424.5cm			2	1.58
4.3	background at 676.5cm			2	1.60
5.1	Light points in dark	400	40	3	1.41
5.2	wave against wharf			3	1.40
5.3	surf at beach			3	1.45
	rear window of moving car			3	1.44
6.1	Two moving light points	400	40	2	1.20
6.2	background at 424.5cm			2	1.28
6.3	background at 676.5cm			2	1.28

was clearly different (Table 2.1). Within Conditions 4, 5, and 6, the effect of distance and pattern of the background upon R was small in general. In Condition 6.1-6.3, K and σ were of the compatible level with those in Condition 1, but $\overline{R}$ was smaller. This is understandable because nothing was visible at the position of Q_1 once the movement of the two Q's had started and, at the end of one cycle of the movement, the two Q's were perceived right in front of the self with the background far away. In Conditions 4 and 5, the values of σ were of the same order with those in Condition 1 but the range of K was between –0.44 and –0.92. These values of K were of the same order with the those in Table 2.1B and were much closer to –1 than the range of K between – 0.29 and –0.31 in Condition 1. This difference in K is reflected to larger values of $\overline{R}$. The value of $\overline{R}$ was largest in the illuminated space (Condition 2). Introducing the background in the dark VS does not affect much to the structure of VS. However, VS under illumination is qualitatively different from dark VS. VS under natural conditions will be discussed in Chapter 4.

2.4. Derivations and Explanations

2.4.1. *Supplementary Explanations to Sec.2.2.1*

Curvature and length of curve

Fig. 2.8A shows a smooth and untwisted curve C in a 3-D Euclidean space. The curve is given by $\mathbf{x}(t) = (x_\alpha(t))$, $\alpha = 1, 2, 3$. The variable t by which points P moves on the curve is called the *parameter*. What is meant by being smooth is that $x_\alpha(t)$ have continuous derivatives up to and including the third order. A twisted curve is said to have torsion. As $\mathbf{x}(t)$ has no torsion, we can define the circle that has a higher degreee of contact with the curve at each point P(t). It means that the first three derivatives of the circle are equal to those of the curve. The circle is called the osculating circle. Denote by O(t) and r(t) its center and radius.

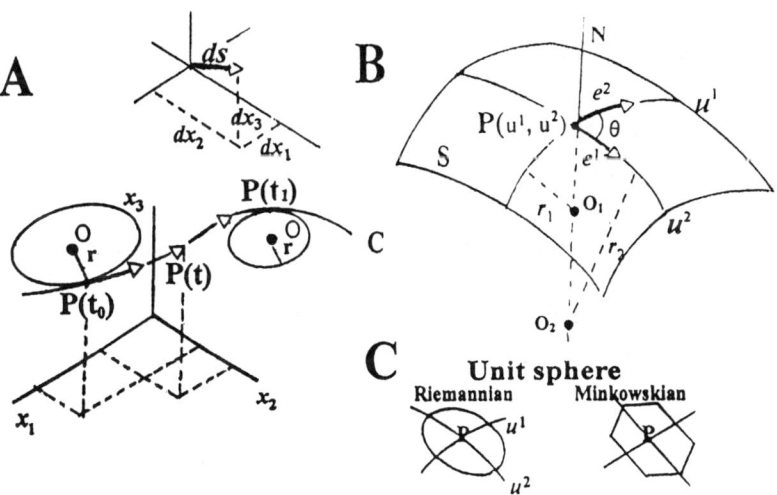

Fig.2.8 Explanation of curvature

The curvature of $\mathbf{x}(t)$ at point P(t) is given by

48 Global Structure of Visual Space

$$\kappa(t) = \frac{1}{r(t)}$$

In Fig.2.8A, $\mathbf{x}(t)$ is less curved at $P(t_0)$ and more curved at $P(t_1)$. Namely $r(t_0) > r(t_1)$ and $\kappa(t_0) < \kappa(t_1)$. If a curve is straight at point $P(t)$, $r(t) = \infty$ and $\kappa(t) = 0$. There is one more difference between $\kappa(t_0)$ and $\kappa(t_1)$; $O(t_0)$ and $O(t_1)$ are on the different sides of the curve. Hence, if sign is assigned to $\kappa(t)$, one can be called positive and the other negative. Which side is to be called positive is arbitrary. Should numerical expression be given to $\kappa(t)$, the value is defined by the inverse of the unit by which $r(t)$ is defined. The tangent line $\mathbf{e}$ to $\mathbf{x}(x_1, x_2, x_3)$ and its length ds are

$$\mathbf{e} = (dx_1, dx_2, dx_3)$$

$$ds = \sqrt{\sum_\alpha dx_\alpha^2}, \quad \alpha = 1, 2, 3 \tag{2.4.1}$$

Since the arc length of a curve segment on C is given by the sum of ds's, ds is called *line element*. By using the parameter t, we can write

$$ds^2 = d\mathbf{e}.d\mathbf{e}, \quad d\mathbf{e} = \frac{d\mathbf{e}(t)}{dt} dt$$

$$ds = \sqrt{\sum_\alpha \left(\frac{dx_\alpha(t)}{dt}\right)^2} \, dt \tag{2.4.2}$$

When t changes from 0 to t, the length of arc, d, is defined as

$$d = \int_0^t \sqrt{\sum_\alpha \left(\frac{dx_\alpha(t)}{dt}\right)^2} \, dt \tag{2.4.3}$$

Surface

Fig.2.8B shows a portion S of a smooth surface in 3-D Euclidean space. A point $P(x_1, x_2, x_3)$ in S can be represented as $x(u^i, u^j)$ by curvilinear coordinates, u^i-constant curve and u^j-constant curve passing through P. It is a matter of convenience to use superscript for u. Suppose a smooth curve C on S. The tangent **e** and the square of line element ds are

$$\mathbf{e} = (u^i), \quad i = 1, 2$$

$$ds^2 = d\mathbf{e} \cdot d\mathbf{e}, \quad d\mathbf{e} = \mathbf{e}_{u1} du^1 + \mathbf{e}_{u2} du^2$$

Hence,

$$ds^2 = \sum_\alpha \left(\frac{dx_\alpha}{dt} dt \right)^2 = \sum_\alpha \left(\frac{\partial x_\alpha}{\partial u^i} \frac{du^i}{dt} dt \right)^2$$

$$= \sum_\alpha \left(\sum_i \sum_j \frac{\partial x_\alpha}{\partial u^i} \frac{\partial x_\alpha}{\partial u^j} du^i du^j \right), \quad i, j = 1, 2$$

Define

$$g_{ij} = \sum_\alpha \frac{\partial x_\alpha}{\partial u^i} \frac{\partial x_\alpha}{\partial u^j}$$

Then,

$$ds^2 = \sum_i \sum_j g_{ij} du^i du^j \tag{2.4.4}$$

This is called the *first fundamental (metric) form* of S. Often the summation symbol is omitted when an index occurs twice, once as a superscript and once as a subscript (Einstein convention).

$$ds = \sqrt{g_{ij} du^i du^j}$$

In Fig.2.8 B, tangent vectors to u^1-curve and u^2-curve, $\mathbf{e}_1$ and $\mathbf{e}_2$, at P forms an angle θ. In the space (x_1, x_2, x_3), g_{12} (=g_{21}) is the scalar (inner) product $\mathbf{e}_1 \mathbf{e}_2$, and g_{11}, g_{22} determines the lengths $|\mathbf{e}_1|$, $|\mathbf{e}_2|$. Namely,

$$|\mathbf{e}_1| = \sqrt{g_{11}}, \quad |\mathbf{e}_2| = \sqrt{g_{22}}, \quad \text{and} \quad \cos \theta = (g_{12} / \sqrt{g_{11} g_{22}}).$$

When P moves along a path C as the parameter t changes from t_1 to $t_2 = t_1 + t$, the length of the path, d_{12}, is defined by

$$d_{12} = \int_0^t \sqrt{g_{ij}(u^i(t), u^j(t)) \frac{du^i}{dt} \frac{du^j}{dt}}\, dt \qquad (2.4.5)$$

Among paths connecting $P(t_1)$ and $P(t_2)$, the one that makes d_{12} shortest is called the *geodesic* between P_1 and P_2 on the surface S. A geodesic corresponds to a straight line on a flat plane, and d_{12} along this path is the distance between the two points. In general, to define distance on a curved surface, two steps are necessary. One is to find out the shortest path and the other is to carry out the integration (2.4.5) on that path. The mathematics to solve this problem is called calculus of variations.

The length of vectors representing ds changes according to direction from P. The trace of these vectors for $ds = 1$ is call the *unit-sphere* around P. When S is a 2-D Riemannian space and ds is defined in Eq.(2.4.4), the unit-sphere is an ellipse (the left in Fig.2.8C). If a more general form of the following property is taken as the unit sphere, the space is called locally Minkowskian. What is shown on the right in C is an example. In this case, what is required for the unit-sphere is that it is convex and symmetric around P. In this case, the distance d_{12} between P_1 and P_2 is defined in a different way from Eq.(2.4.5). If S is locally Minkowskian, it is called a Finsler space which includes Euclidean space as a special case. In this book, only Riemannian space will be dealt with.

Curvature K

As to each point P, we can define a set of u^i-constant curves with the osculating circle (center O_i and radius r_i and the curvature κ_i, $i = 1, 2$). These values are different according to the directions of the set. Then, there are such u^i-constant curves that κ_i is maximum in one direction and minimum in the other direction. These are called *principal directions* at P. In Fig.2.3B, u^1 and u^2 are the principal directions. It can be shown that the angle θ between principal directions is $90°$ (hence $g_{12} = 0$) and O_1 and O_2 are on the same line **N** that is perpendicular to S at P. The *Gaussian curvature* of S at P is defined as

$$K(P) = \kappa_1 \kappa_2 = \frac{1}{r_1}\frac{1}{r_2} \qquad (2.4.6)$$

This is not the only way to define curvature at P. For instance, $H(P) = (\kappa_1 + \kappa_2)/2$ is called the mean curvature. In this book, only K of (2.4.6) will be used. In contrast to H, K is called the *total curvature*, and it can be defined without going back to r's. If we use b_{ij}, the so-called the second fundamental form of S around P, K(P) can be expressed in a simple form

$$K(P) = (b_{11}b_{22} - (b_{12})^2) / (g_{11} g_{22} - (g_{12})^2)$$

Curvature K can be defined in terms of g's alone, but the expression is lengthy even in two-dimensional surface S. The discussion above can be extended to Riemannian space of higher dimension. In this book, however, all equations will be given in two-dimensional surface S, even when 3-D visual space VS^3 is discussed.

Like P in Fig.2.8B, if two u^i-constant curves in the principal directions are curved in the same direction with regard to N, then $K(P) > 0$. If two u^i-constant curves are curved in the opposite direction, O_1 and O_2 are on the different sides of S, one κ is positive and the other κ is negative. Hence, $K(P) < 0$. A surface having negative K(P) looks like a saddleback. For a flat surface, both κ_1 and κ_2 are zero and $K = 0$. However, a flat plane is not the only case in which $K = 0$. The surface of a cylindrical can has $K = 0$. It is a circle in the horizontal direction ($\kappa > 0$) and straight in the vertical direction ($\kappa = 0$), and $K = 0$. This surface is equivalent to a flat surface ($K = 0$) in the sense that figures can be mapped without any distortion from one to the other. This is the reason that we can wrap a cylindrical can by a flat label sheet with preprinted figures. If we want to have a figure on the surface of a ball ($K > 0$), we cannot wrap the ball with a sheet of paper ($K = 0$) on which the figure is preprinted.

2.4.2. *Derivations of Equations in Secs. 2.2.2 and 2.2.3*

Riemannian space R^2 of constant curvature K

52 Global Structure of Visual Space

When K(P) is a constant K for all points P in the space under discussion, the situation becomes simpler. According to whether $K < 0$, $K = 0$, or $K > 0$, the space is *hyperbolic*, *Euclidean*, or *elliptic*. An elliptic surface R^2 ($K > 0$) can be visualized as the surface of sphere Ω of radius r (> 0) in E^3 (Fig.2.9A). A straight line in R^2 is represented by a great circle on the sphere. A meridian circle is a great circle. There are two meridian circles in Fig. 2.9A; L_i passing through P_i and L_j passing through P_j. All meridian circles meet at the north pole N and at the south pole. The figure shows another great circle L_{ij} passing through P_i and P_j. The distance between these points, δ_{ij}, is represented by the arc of this circle. This is the shortest pass from P_i to P_j on the surface Ω. A jet-plane flies along this path. All great circles passing through P_j intersect L_i sooner or later.

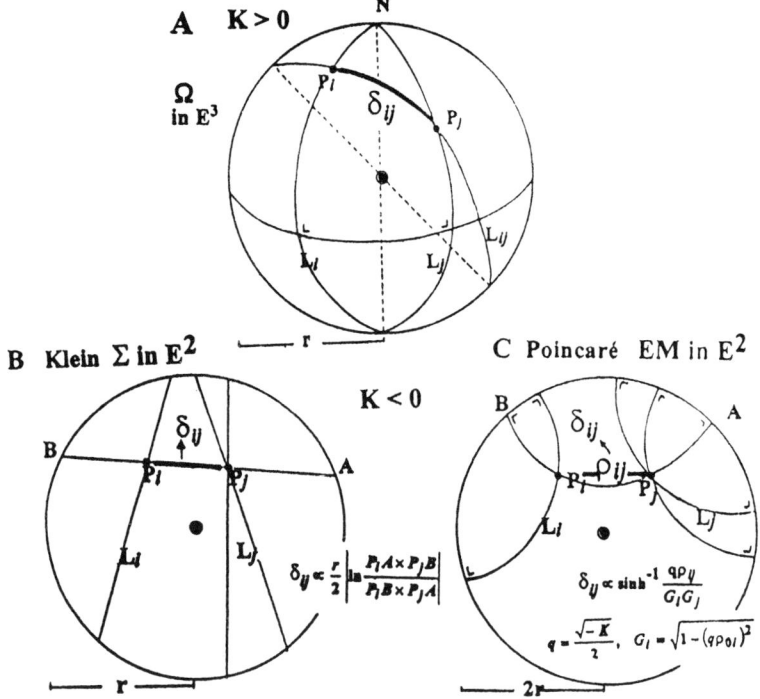

Fig.2.9 Three representations of R^2 of constant K

Namely, in Ω there are no two geodesics that do not intersect. The one denoted as L_j has the longest distance before intersecting L_i.

One way to visualize a portion of hyperbolic surface R^2 ($K < 0$) in E^3 is to take the surface of a pseudosphere or a saddleback. Most textbooks refer to this illustration. However, it is not possible to represent the entire R^2 in this way. There are two ways to represent the entire R^2 of $K < 0$ in a circle on a Euclidean plane E^2. One is given by Klein (Fig.2.9B) and the other by Poincaré (Fig.2.9C). In this book, the former is denoted as Σ and the latter as EM (Euclidean map). In Σ, all points in R^2 of curvature radius r are represented as points P inside a circle of radius r. Points on the circumference represent points at infinity. One advantage of Σ is that a straight line in R^2 is represented by a straight line L in Σ. However, the distance δ_{ij}, between P_i and P_j must be measured in the following way. Denote by A and B the intersections of the straight line connecting P_i and P_j with the circumference, and by P_iA *etc.* the Euclidean distance between P_i and A *etc.* (Fig.2.9B). Then,

$$d_{ij} = \frac{r}{2}|\ln CR|, \quad CR = \frac{P_iA \times P_jB}{P_iB \times P_jA}$$

where ln means the natural logarithm, and CR is called the cross ratio. When $P_i = P_j$, CR $=1$ and $d_{ij} = 0$. When P_i approaches B, lnCR goes $-\infty$ in the current notation. If the notations A and B are exchanged, lnCR goes $+\infty$. In either notation, d_{ij} approaches ∞. As shown in Fig, 2.9B, we can draw many lines L_j's through a point P_j (not on a given line L_i) that do not intersect L_i within the entire space of R^2. This is a characteristic of hyperbolic space.

The Klein's model Σ of hyperbolic space R^2 is explained in many books and hyperbolic trigonometry can be directly derived from Σ (*e.g.*, Busemann and Kelley, 1953). On the other hand, the Poincaré's model EM is not so widely referred to. All points in R^2 of curvature radius r are represented as points P inside a circle of radius 2r in E^2 (Fig.2.9C). Again, points on the circumference represent points at infinity. In EM, a straight line in R^2 is represented by a circle that is perpendicular to the circumference at its both intersections. It will be clear that, in a hyperbolic plane R^2, there are many lines L_i and L_j that do not intersect.

54 *Global Structure of Visual Space*

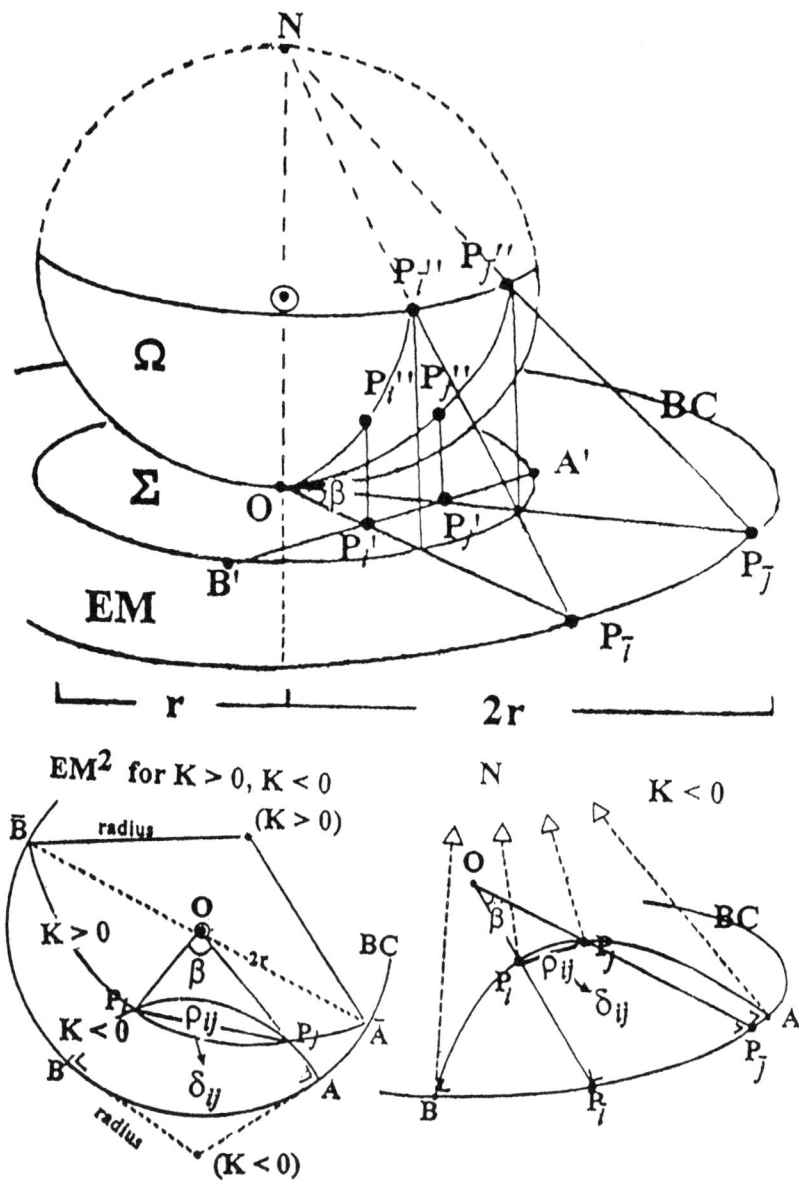

Fig.2.10 Euclidean map EM^2

The distance d_{ij} between P_i and P_j is related to the chord ρ_{ij} of the circle (Eq.2.2.1). All equations in Sec.2.2 were derived in EM. It might be appropriate to explain EM^2 first in general term on the basis of Σ and then to relate it to VS. EM^3 is obtained by rotating EM^2 around O.

Fig.2.10 shows the relationship between Σ and EM^2. Consider a sphere of radius r on Σ. Its south pole is at the center O of Σ. Points P' in Σ are vertically projected as P'' onto the lower hemisphere Ω. The circumference of Σ representing points at infinity is projected to the equator of sphere. By the stereographical projection from the north pole N of sphere, Σ is mapped as a circle of radius 2r in the same plane in which Σ is located. This is EM^2 when $K < 0$. Points P''_i and P''_j on the equator appear on its circumference as P_i and P_j. The circumference is called the basic circle BC in Fig.2.3. Points on it are at infinity. The radial straight line OP'_i in Σ is the great circle OP''_i in Ω and the straight line OP_i in EM^2. Notice that OP_i is perpendicular to BC. To represent VS, only a half of EM is necessary. Consider two points P'_i and P'_j on radial vectors OP_i and OP_j in Σ. The angle between OP_i and OP_j is denoted as β. The straight line connecting P'_i and P'_j intersects the circumference of Σ at A' and B'. These points appear in EM^2 as P_i, P_j and A, B on the circle. The circle is perpendicular to BC at A and B. This part is magnified in the lower right figure. The arc between P_i and P_j represents the straight line connecting these points in R^2 and the distance δ_{ij} is related to the length ρ_{ij} of the chord in EM^2. Eq.(2.2.1) gives the relation between hyperbolic distance δ_{ij} and ρ_{ij}.

One advantage of EM is that the same EM can be used to represent R^2 irrespective of whether $K < 0$, $K = 0$, or $K > 0$. When $K = 0$, EM^2 directly represents R^2 and δ_{ij} in R^2 is equal to ρ_{ij}. When $K > 0$, the lower hemisphere Ω in Fig.2.8A is considered as the model for elliptic space and then projected to EM^2. In Fig.2.9A, a great circle represents a straight line in R^2. The portion in the southern hemisphere of a meridian circle L_i appears in EM^2 as OP_i. In this representation, Σ in Fig. 2.10 does not play any role. When $K < 0$, the sphere Ω is not the model for elliptic space. It is the mediator to change Σ to EM^2. In both cases, $K < 0$ and $K > 0$, points on BC represent points at infinity in R^2, and radial straight lines from O in EM^2 represent radial straight lines in

56 *Global Structure of Visual Space*

R². However, how the straight line connecting P_i and P_j in R^2 is represented in EM^2 (the lower left figure) is different according to whether K < 0 or K > 0. In the former, it is the arc of circle perpendicular to BC. Its center and radius are shown in the figure. In the latter, it is the arc of circle intersecting BC at antipodal points. Denote by $\overline{A}$ and $\overline{B}$ its intersecting points with BC. Then the line connecting $\overline{A}$ and $\overline{B}$ passes through O. Its center and radius are shown in the figure.

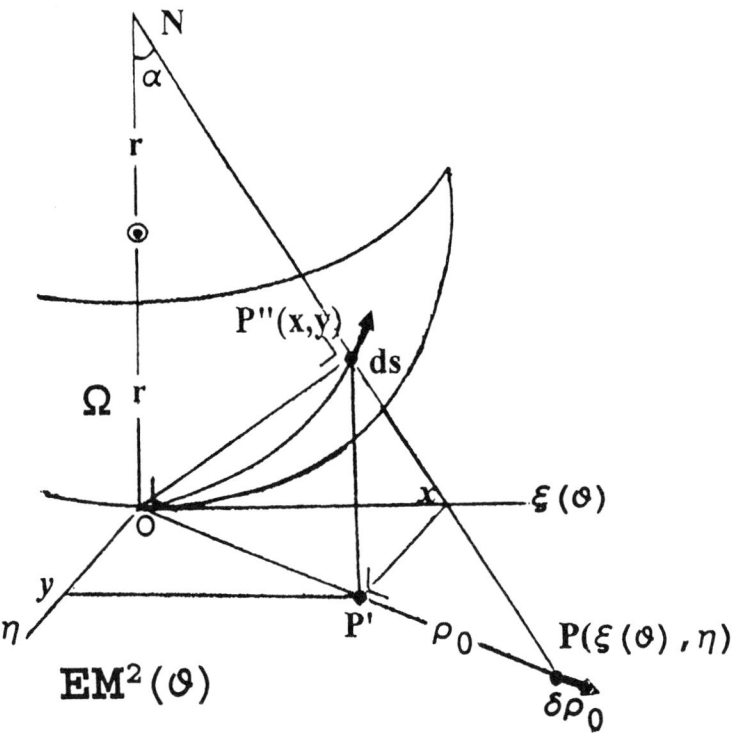

Fig.2.11 *ds* on meridian and $d\rho_0$ in EM

Derivations of Eqs.(2.2.1) and (2.2.3)

Corresponding to Fig.2.3C, Cartesian coordinates $\xi(\theta)$ and η are introduced in Fig.2.11. Now, we are considering EM(θ) for a slanted plane of elevation angle ϑ_V in VS. Point $P(\xi(\vartheta),\eta)$ represents P_V in Fig.2.3B. In both cases, $K < 0$ and $K > 0$, the radial Euclidean distance ρ_0 from O to $P(\xi(\vartheta),\eta)$ in $EM(\theta)$ represents visual radial distance δ_0 in $VS^2(\vartheta_V)$. Eq.(2.2.3) gives the relationship between the two. Similarly, as given in Eq.(2.2.1), ρ_{ij} between P_i and P_j in $EM(\vartheta)$ determines the visual distance δ_{ij}.

One way to derive Eq.(2.2.3) is to relate the line element $d\rho_0$ at $P(\xi(\theta),\eta)$ in $EM(\theta)$ to the corresponding line element ds at $P''(x, y)$ on Ω. Because P'' is on Ω of radius r, the free parameters are x and y only. Using two right triangles, $\Delta NOP''$, ΔNOP (Fig. 2.11), and denoting by α the angle $\angle ONP(P'')$, we have

$$\cos\alpha = \frac{NP''}{2r} = \frac{2r}{NP} \quad \text{and hence} \quad NP'' = \frac{4r^2}{NP}$$

$$\frac{\xi(\theta)}{x} = \frac{\eta}{y} = \frac{NP}{NP''} = \frac{NP^2}{4r^2} \quad \text{and hence} \quad x = \frac{4r^2}{NP}\xi(\theta), \quad y = \frac{4r^2}{NP}\eta$$

$$ds^2 = dx^2 + dy^2 = \frac{4r^2}{NP^2}d\rho_0^2, \qquad d\rho_0^2 = d\xi(\theta)^2 + d\eta^2$$

Since $NP^2 = \rho_0^2 + 4r^2$ from ΔNOP, and $1/r^2$ is $-K$ when VS is hyperbolic and K when VS is elliptic,

$$ds = \left(1 - \frac{K}{4}\rho_0^2\right)^{-1} d\rho_0 \quad \text{when } K < 0$$

$$ds = \left(1 + \frac{K}{4}\rho_0^2\right)^{-1} d\rho_0 \quad \text{when } K > 0.$$

Using the formulae

$$\int \frac{1}{ax^2+b}dx = \frac{1}{\sqrt{|ab|}}\tanh^{-1}\frac{\sqrt{|ab|}}{b}x \quad \text{when } ab < 0 \text{ and } |x| < \sqrt{\left|\frac{b}{a}\right|}$$

$$\int \frac{1}{ax^2+b}dx = \frac{1}{\sqrt{ab}}\tan^{-1}\frac{\sqrt{ab}}{b}x \quad \text{when } ab > 0$$

we have

$$K < 0 \quad \delta_0 = \frac{2}{\sqrt{-K}}\tanh^{-1}(\frac{\sqrt{-K}}{2}\rho_0) \quad \rho_0 < 2r, \ r = 1/|K| \ (|K| < 1),$$

$$K > 0 \quad \delta_0 = \frac{2}{\sqrt{K}}\tan^{-1}(\frac{\sqrt{K}}{2}\rho_0)$$

Because $\sqrt{-K}/2$ or $\sqrt{K}/2$ appears often, it is denoted as q and

$$K < 0, \ \delta_0 = q\tanh^{-1}(q\rho_0), \qquad K > 0, \ \delta_0 = q\tan^{-1}(q\rho_0)$$

These are equivalent to Eq.(2.2.3). Because the straight line from O to $P(\xi(\vartheta),\eta)$ is a geodesic in $EM^2(\vartheta)$, it was easy to define δ_0 by integrating ds in the way stated in Sec.2.4.1.

In order to relate δ_{ij} to ρ_{ij}, Eq.(2.2.1), it is simpler to use the law of cosines.

$K > 0$

$$\cos\left(\frac{\delta_{ij}}{r}\right) = \cos\left(\frac{\delta_{0i}}{r}\right)\cos\left(\frac{\delta_{0j}}{r}\right) + \sin\left(\frac{\delta_{0i}}{r}\right)\sin\left(\frac{\delta_{0j}}{r}\right)\cos\beta \qquad (2.4.7a)$$

$K = 0 \qquad \delta_{ij}^2 = \delta_{0i}^2 + \delta_{0j}^2 - 2\delta_{0i}\delta_{0j}\cos\beta \qquad (2.4.7b)$

$K < 0$
$$\cosh\left(\frac{\delta_{ij}}{r}\right) = \cosh\left(\frac{\delta_{0i}}{r}\right)\cosh\left(\frac{\delta_{0j}}{r}\right) - \sinh\left(\frac{\delta_{0i}}{r}\right)\sinh\left(\frac{\delta_{0j}}{r}\right)\cos\beta$$
(2.4.7c)

The following trigonometric functions of double angle ($2q = 1/r$ in Eq. 2.2.2) are useful in deriving Eq.(2.2.1) and equations of alleys.

$K > 0 \quad \tan\left(\dfrac{\delta_0}{r}\right) = \tan(2q\delta_0) = \dfrac{2\tan q\delta_0}{1 - \tan^2 q\delta_0}$

$K > 0 \quad \tanh\left(\dfrac{\delta_0}{r}\right) = \tanh(2q\delta_0) = \dfrac{2\tan q\delta_0}{1 + \tanh^2 q\delta_0}$

$K > 0 \quad \sin\left(\dfrac{\delta_0}{r}\right) = \sin(2q\delta_0) = \dfrac{2\tan q\delta_0}{1 + \tan^2 q\delta_0}$

$K < 0 \quad \sinh\left(\dfrac{\delta_0}{r}\right) = \sinh(2q\delta_0) = \dfrac{2\tanh q\delta_0}{1 - \tanh^2 q\delta_0}$

$K > 0 \quad \cos\left(\dfrac{\delta_0}{r}\right) = \cos(2q\delta_0) = \dfrac{1 - \tan^2 q\delta_0}{1 + \tan^2 q\delta_0}$

$K < 0 \quad \cosh\left(\dfrac{\delta_0}{r}\right) = \cosh(2q\delta_0) = \dfrac{1 + \tanh^2 q\delta_0}{1 - \tanh^2 q\delta_0}$

Equations when $K < 0$ can be obtained by multipying those when $K > 0$ by $i = \sqrt{-1}$. Using Eq.(2.2.3), these equations can be written in terms of variables in EM.

60 Global Structure of Visual Space

$$K > 0 \qquad \tan\left(\frac{\delta_0}{r}\right) = \tan(2q\delta_0) = \frac{2q\rho_0}{1-(q\rho_0)^2} \qquad (2.4.8a)$$

$$K < 0 \qquad \tanh\left(\frac{\delta_0}{r}\right) = \tanh(2q\delta_0) = \frac{2q\rho_0}{1+(q\rho_0)^2} \qquad (2.4.8b)$$

$$K > 0 \qquad \sin\left(\frac{\delta_0}{r}\right) = \sin(2q\delta_0) = \frac{2q\rho_0}{1+(q\rho_0)^2} \qquad (2.4.8c)$$

$$K < 0 \qquad \sinh\left(\frac{\delta_0}{r}\right) = \sinh(2q\delta_0) = \frac{2q\rho_0}{1-(q\rho_0)^2} \qquad (2.4.8d)$$

$$K > 0 \qquad \cos\left(\frac{\delta_0}{r}\right) = \cos(2q\delta_0) = \frac{1-(q\rho_0)^2}{1+(q\rho_0)^2} \qquad (2.4.8e)$$

$$K < 0 \qquad \cosh\left(\frac{\delta_0}{r}\right) = \cosh(2q\delta_0) = \frac{1+(q\rho_0)^2}{1-(q\rho_0)^2} \qquad (2.4.8f)$$

The left side term of (2.4.7 a, c) is

$$K < 0 \qquad\qquad\qquad K > 0$$

$$\cosh(2q\delta_{ij}) = 1+2\sinh^2(q\delta_{ij}) \qquad \cos(2q\delta_{ij}) = 1-2\sin^2(q\delta_{ij})$$

By using (2.4.8), G's defined in Eq.(2.2.2), and

$$\rho_{ij}^2 = \rho_{0i}^2 + \rho_{0j}^2 - 2\rho_{0i}\rho_{0j}\cos\beta, \qquad \cos\beta = \frac{\rho_{0i}^2 + \rho_{0j}^2 - \rho_{ij}^2}{2\rho_{0i}\rho_{0j}}$$

we can reduce the right side term of (2.4.7 a, c) to

| $K < 0$ | $K > 0$ |

$$1 + 2\frac{(q\rho_{ij})^2}{G_i^2 G_j^2} \qquad\qquad 1 - 2\frac{(q\rho_{ij})^2}{G_i^2 G_j^2}$$

Hence, we have the following form that is equivalent to Eq.(2.2.2)

| $K < 0$ | $K > 0$ |

$$\sinh q\delta_{ij} = \frac{(q\rho_{ij})}{G_i G_j} \qquad\qquad \sin q\delta_{ij} = \frac{(q\rho_{ij})}{G_i G_j}$$

P-and D-alleys and H-curves

Fig.2.12A shows in $EM^2(\vartheta)$ the geodesic curve representing one member of P-alleys passing through $P_1(\rho_{01}\ \eta_1)$, $K < 0$ on the left ($\eta < 0$) and $K > 0$ on the right ($\eta > 0$). The curve is parallel to the axis $\xi(\vartheta)$. When $K < 0$, as explained with regard to Fig.2.9B and C, there are infinite number of geodesic curves (circles perpendicular to BS) that pass through P_1 and do not intersect $\xi(\vartheta)$. Among those geodesic curves, Luneburg selected a particular one that is perpendicular to the η-axis. Since VS and EM are conformal, this geodesic curve captures the impression in VS that both $\xi_V(\vartheta_V)$ and the line passing P_{V1} stretch straight toward infinity. When $K > 0$, as explained with regard to Fig.2.9A, there are no two geodesic curves that do not intersect if the whole area of EM is taken. Such a straight line passing P_{V1} is regarded as the member of P-alley that stretches together with $\xi_V(\vartheta)$ from the η_V-axis that does not meet $\xi_V(\vartheta)$ within BC(ϑ). This is the geodesic curve perpendicular to the η-axis. Define by $\overline{P}$ the point at which the geodesic curve intersects the η-axis, and take a point $P(\rho_0, \eta)$ on this curve. Then, $\Delta PO\overline{P}_p$ is a right triangle, hyperbolic or elliptic, and we can use the following formulae.

62 *Global Structure of Visual Space*

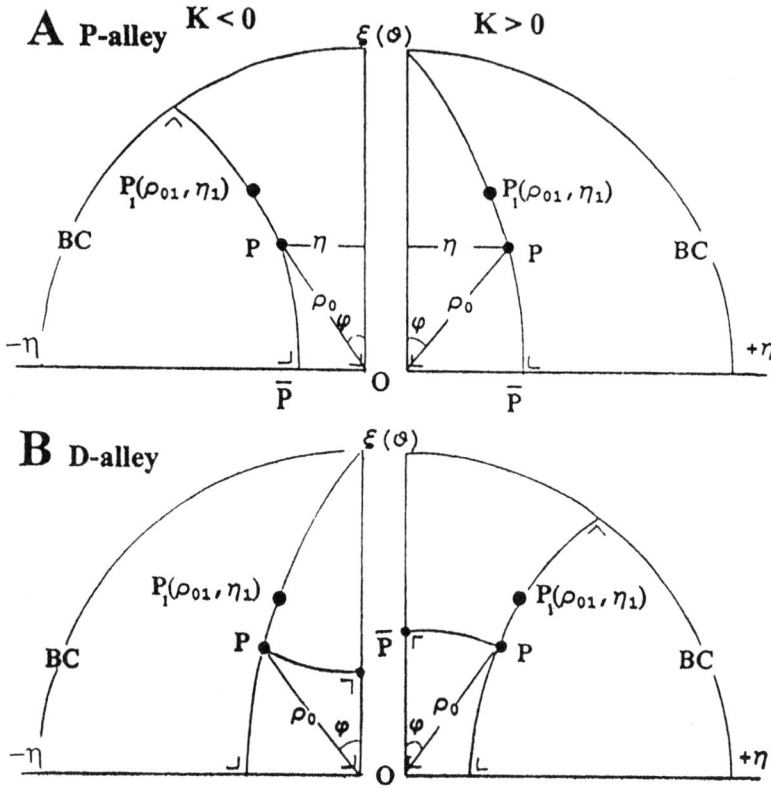

Fig.2.12   P- and D- alleys in $EM^2(\vartheta)$

$$K < 0 \qquad\qquad\qquad K > 0$$

$$\cos\left(\frac{\pi}{2} - \varphi\right) = \sin\varphi = \frac{\tanh\left(\dfrac{\delta_{0\overline{P}}}{r}\right)}{\tanh\left(\dfrac{\delta_0}{r}\right)} \qquad = \frac{\tan\left(\dfrac{\delta_{0\overline{P}}}{r}\right)}{\tan\left(\dfrac{\delta_0}{r}\right)}$$

Denoting

$$N_P = \tanh\left(\frac{\delta_{0\bar{P}}}{r}\right) \qquad N_P = \tan\left(\frac{\delta_{0\bar{P}}}{r}\right)$$

and using (2.4.8), we have

$$\frac{2q\rho_0}{1+(q\rho_0)^2}\sin\varphi = N_P \qquad \frac{2q\rho_0}{1-(q\rho_0)^2}\sin\varphi = N_P$$

$$(q\rho_0)^2 - C_P(q\rho_0)\sin\varphi + 1 = 0 \qquad (q\rho_0)^2 + C_P(q\rho_0)\sin\varphi - 1 = 0$$

$$C_P = \frac{2}{N_P}$$

Since $\rho_0 \sin\varphi = \eta$, Eqs. (2.2.5) and (2.2.6) are obtained. Using the same equation when P is at the fixed point $P_1(\rho_{01}\ \eta_1)$, we can determine the constant C_P.

A circle having its center $q\dot\eta_P$ on $q\eta$-axis with the radius of qr is given by,

$$(q\xi)^2 + (q\eta - q\dot\eta_P)^2 = (qr)^2$$

Eqs. (2.2.5) and (2.2.6) are such circles. For $P_1(\eta_1 > 0)$

$$K < 0 \qquad\qquad K > 0$$

$$C_P = \frac{1+(q\rho_{01})^2}{q\eta_1} > 0 \qquad C_P = \frac{1-(q\rho_{01})^2}{q\eta_1} > 0$$

Center on $q\eta$-axis $\quad q\dot\eta_P = \dfrac{C_P}{2} \qquad\qquad q\dot\eta_P = \dfrac{-C_P}{2}$

64 Global Structure of Visual Space

Radius $\quad qr = \sqrt{(q\dot{\eta}_P)^2 - 1} \qquad\qquad qr = \sqrt{(q\dot{\eta}_P)^2 + 1}$

Fig.2.13 shows the circles for P_1 on the negative side $(-q\dot{\eta}_p)$. When $K < 0$, the circle can be written as

$$(q\xi)^2 + \left(q\eta - \frac{C_P}{2}\right)^2 = \left(\frac{C_P}{2}\right)^2 - 1$$

$$(q\xi)^2 + (q\eta)^2 - C_P(q\eta) + 1 = 0$$

and the first two terms is equal to $(q\rho_0)^2$. Hence, this comes to Eq.(2.2.5). It will be clear from the left side figure that the circle is orthogonal with the BC (radius = 1). The right side figure shows the case when $K > 0$. In this case, the circle and $q\xi(\vartheta)$ meet on the BC(ϑ).

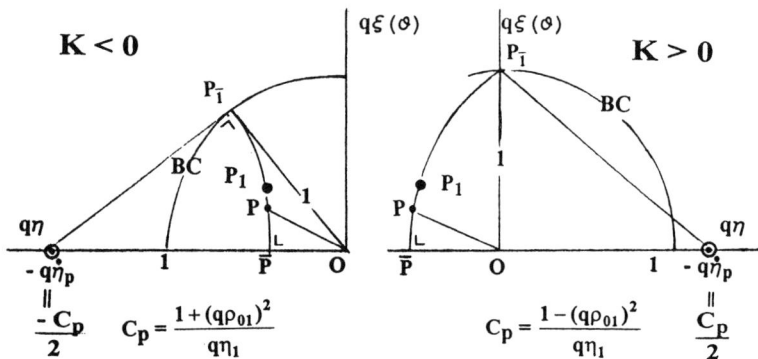

Fig.2.13 P-alley in EM^2 as a circle

Fig.2.12B also shows in EM$^2(\vartheta)$ the curve representing one member of D-alley passing through $P_1(\rho_{01}\ \eta_1)$, $K < 0$ on the left ($\eta < 0$) and $K > 0$ on the right ($\eta > 0$). Take a point P on this curve and consider the geodesic curve passing P that is perpendicular to the $\xi(\vartheta)$. Denote the intersecting point by $\overline{P}$. The curve should satisfy the condition that the perceptual distance $\delta_{P\bar{P}}$ corresponding the arc $P\bar{P}$ is constant for any point P. Then, $\Delta PO\bar{P}$ is a right triangle, hyperbolic or elliptic, and we can use the following formulae.

$K < 0$

$$\sin\varphi = \frac{\sinh\left(\dfrac{\delta_0}{r}\right)}{\sinh\left(\dfrac{\delta_{P\bar{P}}}{r}\right)} = \frac{N_D}{\sinh\left(\dfrac{\delta_{P\bar{P}}}{r}\right)}$$

$K > 0$

$$\sin\varphi = \frac{\sin\left(\dfrac{\delta_0}{r}\right)}{\sin\left(\dfrac{\delta_{P\bar{P}}}{r}\right)} = \frac{N_D}{\sin\left(\dfrac{\delta_{P\bar{P}}}{r}\right)}$$

Using Eqs.(2.4.8), we have

$$\frac{2q\rho_0}{1-(q\rho_0)^2}\sin\varphi = N_D \qquad \frac{2q\rho_0}{1+(q\rho_0)^2}\sin\varphi = N_D$$

$$(q\rho_0)^2 + C_D\rho_0\sin\varphi - 1 = 0 \qquad (q\rho_0)^2 - C_D\rho_0\sin\varphi + 1 = 0$$

Thus, Eqs. (2.2.7) and (2.2.8) are derived. Using the same equation when P is at $P_1(\rho_{01}\,\eta_1)$, we can determine the constant C_D. The curve is an arc of the circle that is perpendicular to the η-axis. For $P_1(\eta_1 > 0)$

$K > 0$

$$C_D = \frac{1-(q\rho_{01})^2}{q\eta_1} > 0$$

$K > 0$

$$C_D = \frac{1+(q\rho_{01})^2}{q\eta_1} > 0$$

Center on $q\eta$-axis $\quad q\dot\eta_D = \dfrac{-C_D}{2} \qquad\qquad q\dot\eta_D = \dfrac{C_D}{2}$

Radius $\quad qr = \sqrt{(q\dot\eta_D)^2 + 1} \qquad\qquad qr = \sqrt{(q\dot\eta_D)^2 - 1}$

Notice that this circle is not a geodesic circle. It is simply the trace of point P satisfying the condition stated above.

An H_α-geodesic curve passing through a fixed point $P_\alpha(\xi_\alpha(\vartheta)>0)$ on the $\xi(\vartheta)$, $\xi_\alpha(\vartheta) = \rho_{0\alpha}$, represents a straight line parallel to the η_V-axis in a plane $VS^2(\vartheta_V)$. Hence, its equation is obtained from the equation of P-alley if η is replaced by $\xi(\vartheta)$ and ρ_{01} by $\rho_{0\alpha}$.

	$K < 0$	$K > 0$
	$C_H = \dfrac{1+(q\rho_{0\alpha})^2}{q\rho_{0\alpha}} > 0$	$C_H = \dfrac{1-(q\rho_{0\alpha})^2}{q\rho_{0\alpha}} > 0$
Center on $q\xi(\vartheta)$-axis	$q\dot\xi_H = \dfrac{C_{H\alpha}}{2}$	$q\dot\xi_H = \dfrac{-C_{H\alpha}}{2}$
Radius	$qr_\alpha = \sqrt{(q\dot\xi_H)^2 - 1}$	$qr_\alpha = \sqrt{(q\dot\xi_H)^2 + 1}$

3. Two Extensions of Luneburg Model

In our daily life, we will never come across two perceptually parallel lines epitomized by P-alley discussed in Chapter 2. Often we see two physically straight and parallel lines, *e.g.*, railway tracks. These are perceptually not parallel, however. For the observer standing in the middle of the tracks, the two lines appear to converge at the horizon. It is called the vanishing point in linear perspective. On the other hand, what we see in a P-alley (Fig.2.3) are parts of lines, perceptually straight and parallel up to infinity, that stretch on a DP plane from the η_V-axis passing through the self. As a matter of course, we see lines that appear straight and parallel under natural conditions. Those are not in this direction. Frames of a window are physically straight and parallel, horizontally and vertically. To our casual observation, they appear in that way on a plane that is more or less perpendicular to the line of sight. P- and D-alleys in this plane were not discussed by Luneburg. These alleys that are not extending from the self will be dealt with in Sec.3.1, theoretically and experimentally.

As pointed out in Sec.2.3.1, the model discussed in Chapter 2 is under the constraint of simple and rigid mapping functions. This is the reason that all visual patterns dealt with in Chapter 2 are in frameless spaces. If theoretical analysis in EM can be related to configuration $\{Q_i\}$ in the physical space with no intervention of mapping functions, we can approach the geometry of VS under more natural conditions. Such an approach becomes possible if data based on assessment of perceived distances are incorporated. Through this approach it will become possible to deal with geometrical patterns we see in our ordinary life. The problems of how to obtain quantitative values representing perceived distances and how to make use of these data will be discussed in Sec.3.2.

3.1. Alleys on a Frontoparallel Plane

3.1.1. *Theoretical Equations*

Most visual patterns we see are on frontoparallel planes. A plane appearing to stand parallel to the plane spanned by the η_V- and ζ_V- axes is denoted as an HP. If a perceptual pattern in this plane is structured according to Riemannian geometry of constant K ($\neq 0$), P- and D-alleys on an HP must exhibit the discrepancy as they do on a plane, DP, stretching from the self. In order to represent in EM^3 an HP and alleys on it, we have to extend the discussion in Chapter 2, because what were dealt with therein were limited to alleys on a horizontal DP ($\theta = 0$) or a slanted DP($\theta \neq 0$) (see Fig.2.3C). Some extensions are straightforward and some are not.

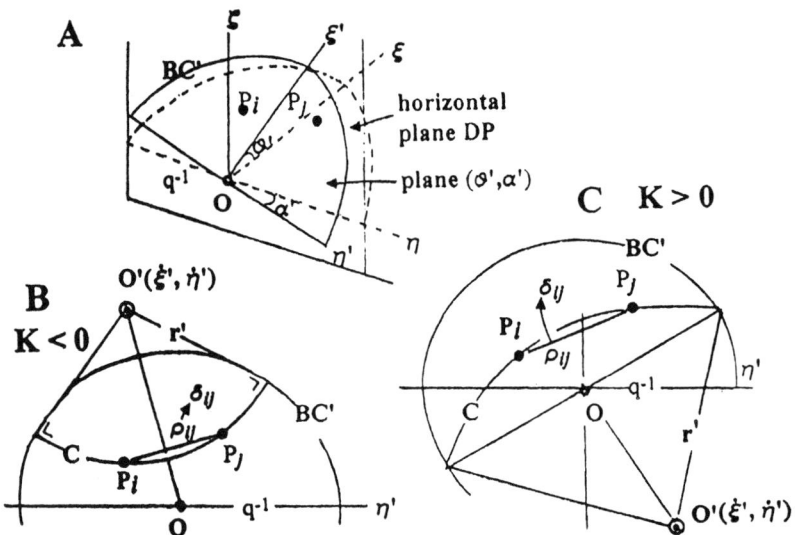

Fig.3.1 A plane in EM^3 defined by P_i and P_j where $\vartheta_i \neq \vartheta_j$

The relation between Euclidean distance ρ_{ij} in $EM^2(\vartheta)$ and the length of visual distance δ_{ij} is given in Eq.(2.2.1). In that context, two points P_i and P_j are defined on the same plane $DP(\vartheta)$. However, the same equations hold even if ϑ_i of P_i and ϑ_j of P_j are different. Take two points in EM^3, $P_i(\xi_i, \eta_i, \zeta_i)$ and $P_j(\xi_j, \eta_j, \zeta_j)$, and denote the plane defined by P_i, P_j, and O as (ϑ', α'). As shown in Fig.3.1A, the ξ- and η-axes as well as BC in the horizontal DP ($\zeta = 0$) are rotated to this plane (ϑ', α') as the ξ'- and η'-axes and BC'. The coordinates of Fig.3.1 are ξ etc, not $q\xi$ etc., and hence the radius of BC and that of BC' are $1/q$. In this plane, we can define the circle C passing through P_i and P_j that is perpendicular to BC' when $K < 0$ (Fig.3.1B) or that meets BC' at antipodal points when $K > 0$ (Fig.3.1C). Define the center of circle C as $O'(\dot{\xi}', \dot{\eta}')$, and its radius as r'. As to how to obtain $O'(\dot{\xi}', \dot{\eta}')$ and r' form $P_i(\xi_i, \eta_i, \zeta_i)$ and $P_j(\xi_j, \eta_j, \zeta_j)$, consult Indow (1988). It is not necessary in this context. The center O' is on the positive side of ξ' when $K < 0$ and on the negative side of ξ' when $K > 0$. The same relation holds between δ_{ij} and chord ρ_{ij} on plane (ϑ', α') as that between δ_{ij} and chord ρ_{ij} on $DP(\vartheta)$ (Eq. 2.2.1). The only difference is that the Euclidean distances, $\rho_{ij}, \rho_{0i}, \rho_{0j}$, have to be defined in the 3-D space (ξ, η, ζ). Namely,

$K < 0$ $\qquad\qquad\qquad\qquad$ $K > 0$

$$\delta_{ij} = \frac{1}{q}\sinh^{-1}\frac{q\rho_{ij}}{G_i G_j} \qquad\qquad \delta_{ij} = \frac{1}{q}\sin^{-1}\frac{q\rho_{ij}}{G_i G_j} \qquad (3.1.1)$$

$$G_i = \sqrt{1-(q\rho_{0i})^2} \qquad\qquad G_i = \sqrt{1+(q\rho_{0i})^2}$$

$$G_j = \sqrt{1-(q\rho_{0j})^2} \qquad\qquad G_j = \sqrt{1+(q\rho_{0j})^2}$$

$$q = \frac{\sqrt{-K}}{2} \qquad\qquad\qquad q = \frac{\sqrt{K}}{2}$$

$$\rho_{ij} = \sqrt{(\xi_i - \xi_j)^2 + (\eta_i - \eta_j)^2 + (\zeta_i - \zeta_j)^2}$$

$$\rho_{0i} = \sqrt{\xi_i^2 + \eta_i^2 + \zeta_i^2} \qquad\qquad \rho_{0j} = \sqrt{\xi_j^2 + \eta_j^2 + \zeta_j^2}$$

These Equations are given by Luneburg (Eq.6.1 in 1950).

First, we have to define the frontoparallel plane passing through point P_1 in the horizontal DP ($\zeta = 0$). Fig.3.2 shows a half of the H-curve in this DP that passes through a fixed point $P_1(\xi_1, 0, 0)$ on the ξ-axis, when $K < 0$ on the left side and when $K > 0$ on the right side. Hereafter, this curve is denoted as $Hh(P_1)$, its center and radius as $O(P_1) = (\dot{\xi}_1, 0, 0)$ and r_1. The two points at which $Hh(P_1)$ intersects BC, the basic circle of this DP, are denoted as $\overline{P}_1$. A point $P(\rho_0, \varphi)$ moving on $Hh(P_1)$ from $\overline{P}_1$ to $\overline{P}_1$ through P_1 represents the perceived P_V that moves, from left to right, on the frontoparallel straight line in the horizontal DP($\vartheta_V = 0$) in VS^3.

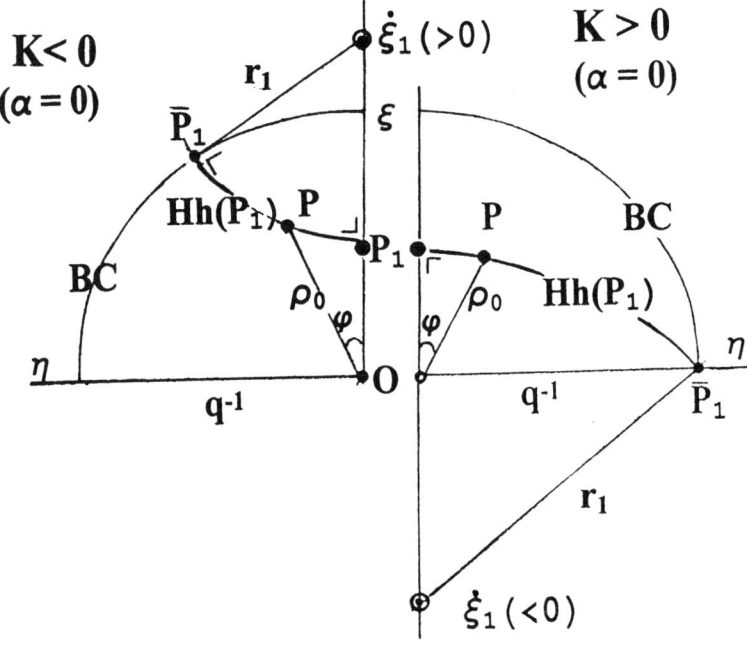

Fig.3.2 H-curve, $Hh(P_1)$, passing through P_1 in DP ($\vartheta = 0$)

According to Eqs.(2.2.9') and (2.2.10')

$K < 0$	$K > 0$	
$\dot{\xi}_1 = \dfrac{C_1}{2q}$	$\dot{\xi}_1 = \dfrac{-C_1}{2q}$	(3.1.2a)
$C_1 = \dfrac{1 + (q\dot{\xi}_1)^2}{q\dot{\xi}_1}$	$C_1 = \dfrac{1 - (q\dot{\xi}_1)^2}{q\dot{\xi}_1}$	
$r_1 = \dfrac{1}{q}\sqrt{(q\dot{\xi}_1)^2 - 1}$	$r_1 = \dfrac{1}{q}\sqrt{(q\dot{\xi}_1)^2 + 1}$	(3.1.2b)
$q = \dfrac{\sqrt{-K}}{2}$	$q = \dfrac{\sqrt{K}}{2}$	

If Fig.3.2 is rotated around the ξ-axis (α in Fig.3.1A = 0 to 90°) so that the η-axis becomes the ζ-axis, we have a part of the sphere in EM^3 that represents the frontoparallel plane in VS^3 passing through P_{V1}. The part of the sphere inside BC is denoted as $HP(P_1)$. When $K < 0$, $HP(P_1)$ is the outer surface of a ball having the radius r_1 that is placed at $\dot{\xi}_1\ (> 0)$. When $K > 0$, $HP(P_1)$ is the inner surface of a ball of radius r_1 placed at $\dot{\xi}_1\ (< 0)$. Point $P(\xi, \eta, \zeta)$ on $HP(P_1)$ satisfies the condition

$$(\xi - \dot{\xi}_1)^2 + \eta^2 + \zeta^2 = r_1 \qquad (3.1.3)$$

$$\rho_0 = \sqrt{\xi^2 + \eta^2 + \zeta^2}, \qquad \xi_1 \leq \rho_0 \leq 1/q$$

Fig.3.3A shows $HP(P_1)$ when $K < 0$. $Hh(P_1)$ is the intersection of $HP(P_1)$ and the plane spanned by the ξ- and η-axes before the rotation. The meridian, $Hh(P_1)$ after the rotation, is denoted as $Hv(P_1)$, because it represents the vertical frontoparallel line passing through P_{V1} in VS^3 $Hv(P_1)$ is the intersection of $HP(P_1)$ and the plane spanned by the ξ- and ζ-axes before the rotation.

72 *Global Structure of Visual Space*

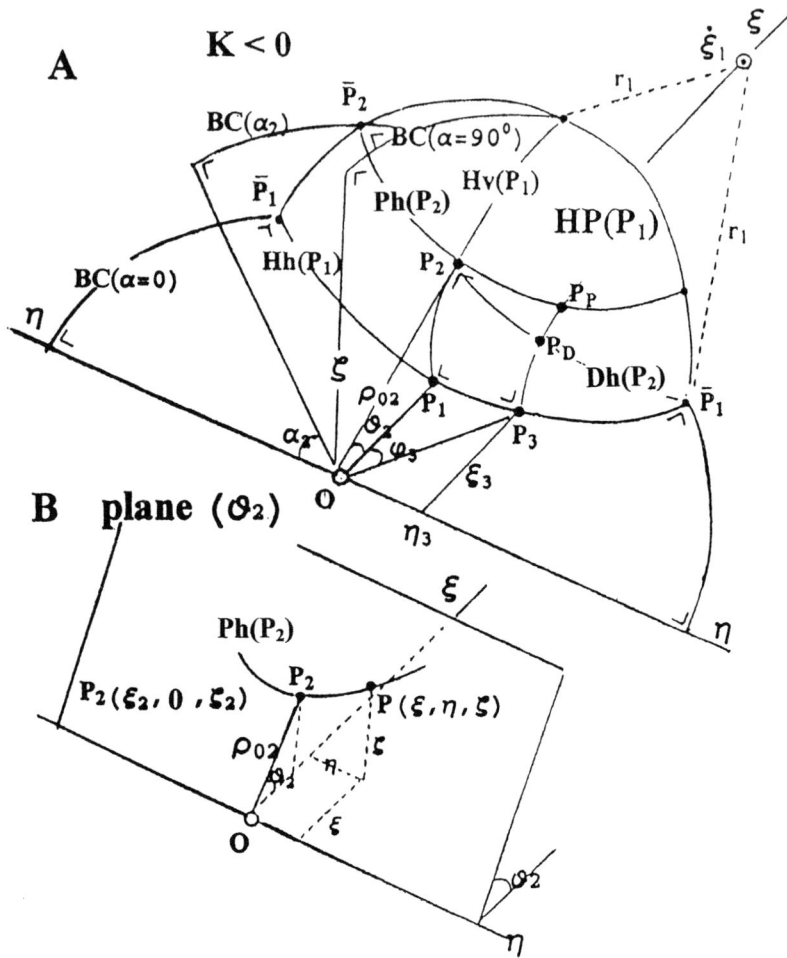

Fig.3.3 Sphere representing a frontoparallel plane in X^3 ($K < 0$)

Let us consider on $HP(P_1)$ a geodesic that is horizontally parallel to $Hh(P_1)$. Take a point P_2 on $Hv(P_1)$ and denote its coordinates as (ξ_2, 0, ζ_2) or (ρ_{02}, 0, ϑ_2). Then, among geodesic curves on $HP(P_1)$ that passes through P_2, we must define one that forms the horizontal P-alley with regard to Hh(P$_1$). The curve is denoted as Ph(P$_2$). Namely, Ph(P$_2$) and Hh(P$_1$) represent two straight lines that are parallel in the horizontal direction on HP(P$_1$). Ph(P$_2$) is given by the intersection of HP(P$_1$) and plane(ϑ_2) spanned by P$_2$ and the η-axis. When K < 0, both HP(P$_1$) and Ph(P$_2$) are convex toward O. When K > 0, both HP(P$_1$) and Ph(P$_2$) are concave. As shown in Fig.3.3B, for a point P(ξ, η, ζ) on Ph(P$_2$)

$$\zeta = a_2\,\xi, \qquad a_2 = \tan\vartheta_2 = \frac{\zeta_2}{\xi_2} \qquad (3.1.4a)$$

Take a point P$_3$ on Hh(P$_1$) and denote its coordinate as (ξ_3, η_3, 0) and (ρ_{03}, φ_3, 0), and denote as H$_V$(P$_3$) the vertical parallel alley passing through P$_3$. For a point P(ξ, η, ζ) on H$_V$(P$_3$),

$$\eta = a_3\,\xi \qquad a_3 = \tan\varphi_3 = \frac{\eta_3}{\xi_3} \qquad (3.1.4b)$$

Define as P$_P$(ξ_P, η_P, ζ_P) the intersection of Ph(P$_2$) and Hv(P$_3$) (Fig.3.3A). Then, P$_P$ must satisfy (3.1.4a and b). From Eq.(3.1.3), we have

$$K<0 \qquad (q\xi_P)^2(1+a_2^2+a_3^2) - 2(q\dot\xi_1)(q\xi_P)+1 = 0$$

$$K>0 \qquad (q\xi_P)^2(1+a_2^2+a_3^2) - 2(q\dot\xi_1)(q\xi_P)-1 = 0$$

Solving this quadratic equation, we have ξ_P, and then η_P and ζ_P

$$K<0 \qquad \xi_P = \frac{q\dot\xi_1 - \sqrt{(q\dot\xi_1)^2 - (1+a_2^2+a_3^2)}}{q(1+a_2^2+a_3^2)} \qquad (3.1.5a)$$

$$K>0 \qquad \xi_P = \frac{q\dot\xi_1 + \sqrt{(q\dot\xi_1)^2 + (1+a_2^2+a_3^2)}}{q(1+a_2^2+a_3^2)} \qquad (3.1.5b)$$

$$\eta_P = a_3\,\xi_P, \qquad \zeta_P = a_2\,\xi_P$$

74 Global Structure of Visual Space

The curve Ph(P$_2$) is perpendicular to the H$_V$(P$_1$) at P$_2$. Furthermore it is perpendicular to BC sphere at $\overline{P}_2$ when K < 0. This point $\overline{P}_2$ is the intersection of HP(P$_1$) and BC(α_2)-curve in the plane rotated with the angle α_2 around the ξ-axis. The perpendicularity is explained in Indow (1988). When K > 0, Ph(P$_2$) and Hh(P$_1$) meet at $\overline{P}_1$ on the η-axis.

The horizontal distance alley Dh(P$_2$) passing through P$_2$ on HP(P$_1$) is the locus of points P$_D$ that have the constant vertical distance from P$_3$ when P$_3$ moves on Hh(P$_1$). Namely $\delta_{3D} = \delta_{12}$. Hence, ($\xi_D$, η_D, ζ_D) of P$_D$ are obtained as functions of P$_3$ (ξ_3, η_3, ζ_3) (Indow, 1988). From (3.1.1),

K < 0 $\qquad\qquad\qquad$ K > 0

$$\sinh^{-1}\frac{q\rho_{3D}}{G_3 G_D} = \sinh^{-1} N_{12} \qquad \sin^{-1}\frac{q\rho_{3D}}{G_3 G_D} = \sin^{-1} N_{12}$$

$$\rho_{3D} = \sqrt{(\xi_D - \xi_3)^2 + (\eta_D - \eta_3)^2 + \zeta_D^2},$$

$$N_{12} = \frac{q\rho_{12}}{G_1 G_2}$$

Hence,

$$(q\rho_{3D})^2 = (N_{12} G_3)^2 G_D^2$$

Because P$_D$ is on HP(P3), it satisfies the equation

$$(\xi_D - \dot\xi_1)^2 + (a_3 \xi_D)^2 + \zeta_D^2 = r_1^2$$

$$a_3 = \pm\frac{\eta_D}{\xi_D}$$

Using these equations, we have

$K < 0$
$$\xi_D = \frac{G_3^2(2N_{12}^2 + 1)}{2q[\{1 + (N_{12}G_3)^2\}q\dot{\xi}_1 - (1 + a_3^2)q\xi_3]} \quad (3.1.6a)$$

$K > 0$
$$\xi_D = \frac{G_3^2(2N_{12}^2 - 1)}{2q[\{1 - (N_{12}G_3)^2\}q\dot{\xi}_1 - (1 + a_3^2)q\xi_3]} \quad (3.1.6b)$$

$$\eta_D = a_3 \xi_D, \quad \zeta_D = \sqrt{r_1^2 - (\dot{\xi}_D - \dot{\xi}_1)^2 - \eta_D^2}$$

All equations above are in EM^3 and have only one parameter q (= $0.5\sqrt{-K}$ or $0.5\sqrt{K}$). To define the curves in the physical space X^3, these must be projected to X^3 and P(ξ, η, ζ) must be transformed to Q(x, y, z). One way of doing so is to use the Luneburg's mapping functions explained in Sec.2.3.1. In this case, one more parameter σ is necessary.

Table 1 in Indow (1988) shows how Q_P and Q_D are different in standard sizes of stimulus point configuration $\{Q_i\}$ and with standard values of K and σ. Discrepancies are experimentally detectable sizes. For example, when Q_1(300, 0, 0) and Q_2(329, 0, 99.9) in cm and K = −0.3, σ = 11.71, (z_P −z_D) is 2.0 cm for Q_3 of ϕ = 13.6°, and 8.4 cm for Q_3 of ϕ = 24.9°. Hence, two series of experiments were perfomed (Indow and Watanabe, 1984b, 1988). The first one is based upon the equations for Ph(P_2) and Dh(P_2) proposed by Indow (1979). Then, Watanabe noticed a problem and Indow changed the equations as described in this section. Besides, the alleys in the first experiment are smaller. Hence, some representative results from the second experiment will be shown in the next section.

3.1.2. *Experimental Results*

Fifteen stimulus points Q_i, five in each of U, M, and L horizontal series, were used (Fig.3.4). Each series was supported by the scaffold. The experiment was conducted in a dark room and Q_i were small light points of 1 mm diameter. The luminance was adjusted according to its position so as to appear with the same brightness. The subject S could move Q_i on the small base in three directions, x, y, z. Displacements of bases in the x- and y-directions on the scaffold were made by pulling strings. On each base, Q_i was fixed to a small jack that was movable up and down by a small motor attached to the base. The subject could move Q_i in the z-direction by means of a toggle switch. The series M was the horizontal DP and M3 was fixed at $(x_{M3}, 0, 0)$. Positions of M3 and furthest left and right Q's were selected so that the subject had the view of the largest possible configuration of points. The head was fixed, but the S was encouraged to scan the configuration by moving both eyes. Because of the scaffold, the experiment was performed in a dark room only.

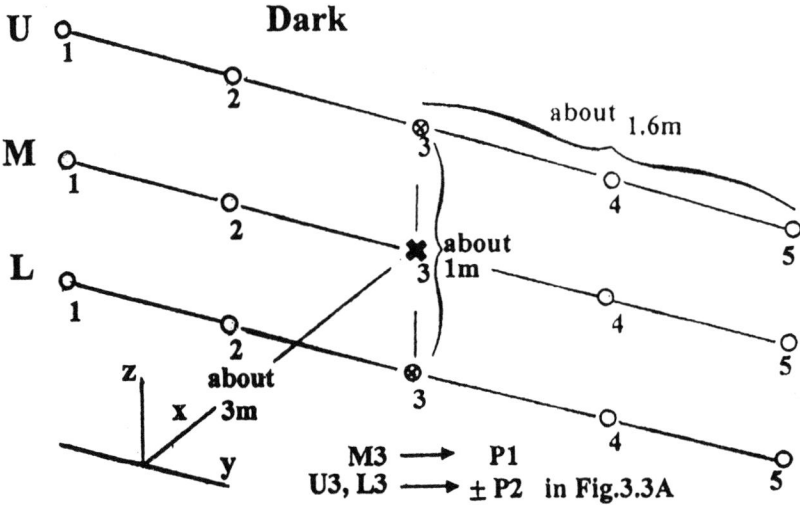

Fg.3.4 Alleys on a frontoparallel plane

First, positions of Q_i in M series, except M3, were adjusted so as to appear frontoparallel and equi-distant, i.e., $\delta_{M1M2} = \delta_{M2M3} = \delta_{M3M4} = \delta_{M4M5}$. The results were denoted as Hh(M3). Then, U3 and L3 were adjusted so that (U3, M3, L3) form the Hv(M3) and $\delta_{M3U3} = \delta_{M3L3}$. Positions of remaining Q_i in U and L series were adjusted so as to meet the following conditions. All appear on the frontoparallel plane, each (UX, MX, LX) appears vertically straight, and either series U and L appear horizontally parallel to the M series to give Ph(U3) and Ph(L3) or UX and LX appear vertically equi-distant from MX to give Dh(U3) and

Dh(L3), X = 1, 2, 4, and 5. Namely, M3 plays the role of P_1 in Fig.3.3, whereas U3, L3 play the roles of $\pm P_2$ and MX the roles of $\pm P_3$.

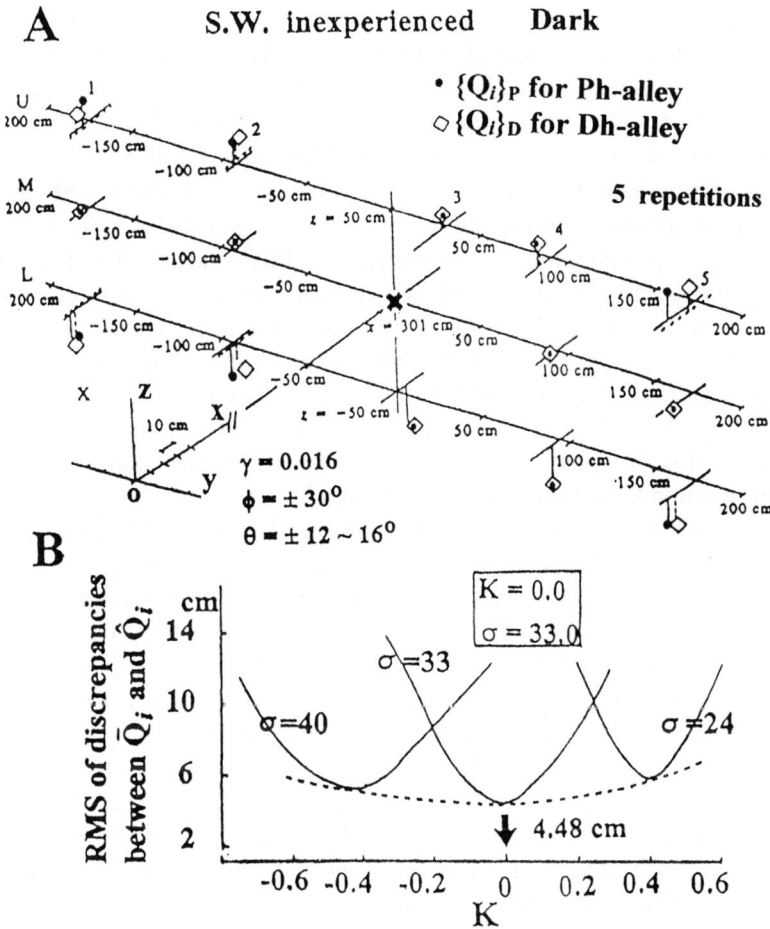

Fig.3.5 An example of P- and D- alleys on a frontoparallel plane

The procedure to construct these horizontal P-alleys and D-alleys on the HP(M3) was explained in detail in Indow and Watanabe (1988). Three S's participated in this experiment. In addition to the standard procedure (Sec.2.1.2), alleys through U3, L3 were also constructed by another procedure (adding Q_i) described in that section. Always, once all the adjustments had been completed, the whole set of Q's were presented. If the S felt it necessary, readjustment of some points, except M3, was allowed. Each S repeated the settings five times over five days and the average result was defined as $\{Q_i\}_X$, X = P or D. Alleys of moving points were obtained from the two other subjects. Herein, only $\{Q_i\}$'s of an inexperienced S by the standard procedure are shown, because the results were essentially the same in all the other cases.

Fig.3.5A shows $\{Q_i\}_P$ for Ph-alleys (black dots) and $\{Q_i\}_D$ for Dh-alleys (white diamonds). The configuration $\{Q_i\}$ spans about $\pm 30^O$ in the horizontal direction and about $\pm 12 \sim 16^O$ in the vertical direction. The subject SW participated in experiments of this kind for the first time. Two configurations, $\{Q_i\}_P$ and $\{Q_i\}_D$, coincide almost perfectly. The coincidence is still closer in the other two experienced S's. This coincidence was observed in the first experiment (1984) also. Namely, visual patterns in a frontoparallel subspace VS^2 seem to follow Euclidean geometry. According to the test calculation stated at the end of Sec. 3.3.1, if this HP plane is hyperbolic as in DP planes, black dots and white diamonds should exhibit discrepancies of observable sizes in the direction of z-axis, especially at U1, L1 and U5, L5. There is no doubt that $\{Q_i\}_P$ and $\{Q_i\}_D$ coincide on HP(M3). Concluding K = 0 on the basis of this coincidence does not presuppose any particular form of mapping between VS and X, except that the same mapping functions hold for $\{Q_i\}_P$ and $\{Q_i\}_D$. On the horizontal DP(θ = 0), Hh(Q_{M3}) is convex when Q_{M3} on the x-axis is far and concave when Q_{M3} is near irrespective of the sign of K (Fig.2.1). The sign of K only affects the inflection distance at which Hh(Q) changes from convexity to concavity (Sec.2.2.3). In Fig.3.5A, the three horizontal series, $\{Q_i\}_M$, $\{Q_i\}_L$, and $\{Q_i\}_U$, are more or less concave along the x-axis toward the S. We do not have much empirical information as to how vertical frontoparallel series Hv(Q) of Q behaves when the distance to Q in the DP is changed or Q is moved from 0 to $\pm \phi^O$.

If we assume the Luneburg's mapping functions, we can project the curves (3.1.5) and (3.1.6) in EM^3 to X^3. Let us denote the configurations as $\{\hat{Q}_i\}_P$ and $\{\hat{Q}_i\}_D$. The values of parameters, K and σ, can be determined so that $\{\hat{Q}_i\}_P$, $\{\hat{Q}_i\}_D$ are closest to data $\{Q_i\}_P$ and $\{Q_i\}_D$. The procedure to estimate the optimum values of K and σ is explained in

Indow and Watanabe (1988). As Q's were symmetrized with regard to the *x*-axis in Fig.2.1, Q's in this experiment were symmetrized with regard to the M-series and to the meridian plane. Let us denote the results as $Q(\bar{x}_i, \bar{y}_i, \bar{z}_i)$. If a set of values (K, σ) is given, theoretical positions $\hat{Q}(\hat{x}_i, \hat{y}_i, \hat{z}_i)$ for $Q(\bar{x}_i, \bar{y}_i, \bar{z}_i)$ in X^3 are determined through Eqs.(3.1.5) and (3.1.6). The degree of coincidence between $\bar{Q}_i$ and $\hat{Q}_i$ was evaluated by RMS (root-mean-squares) of discrepancies in arguments including both Ph- and Dh-alleys. Values of K and σ were systematically varied. Let us denote by min RMS(K|σ) the smallest RMS when K was varied with a fixed value of σ. Then, another value of σ was fixed and K was systematically varied to find min RMS(K|σ). In this way, the set of (K, σ) yielding the smallest min RMS(K|σ) was adopted. Three out of the five curves of min RMS(K|σ) for S.W. are shown in Fig.3.5B. The set (K = 0, σ = 33) gave the best fit. The change of min RMS(K|σ) according to σ was rather small. The situation was the same in the other two S's (Fig. 2 in Indow and Watanabe, 1988).

The goodness of fit in this case may be more affected by the applicability of the Luneburg's mapping functions. The coincidence between $\{Q_i\}_P$ and $\{Q_i\}_D$ in Fig. 3.5A can be regarded as sufficient evidence for that K = 0 in a frontoparallel plane HP. On the other hand, alley experiments in DP's (θ = 0 or > 0) show that K < 0 in these planes (Fig.2.1). We have to conclude that visual patterns are structured in geometrically different ways between these two subspaces in VS^3, DP and HP. That this fact is not necessarily to be taken as a contradiction will be discussed in Sec.5.4.

3.2. Direct Mapping according to Riemannian Metric

In all the experiments hitherto described, the subject S creates a stimulus pattern $\{Q_i\}$ in the physical space X so that it appears in VS as $\{P_{Vi}\}$ that has a specified visual structure. The theoretical pattern $\{P_i\}$ satisfying the structure is defined in EM. In order to fit $\{P_i\}$ in EM to the data $\{Q_i\}$ in X, the Luneburg's mapping functions between X and EM are used.

By these functions, {P_i} in EM is projected to X or {Q_i} in X is projected to EM. The Luneburg's mapping functions are very rigid and inflexible, which severely constrains the applicability of this approach. If the pattern {P_i} representing the visual pattern {P_{Vi}} generated by {Q_i} is directly defined in EM with no intervention of the mapping functions, then the theoretical equation is fitted to this {P_i} in EM. Should the correspondence between X and EM be made explicit, it can be obtained by comparing {P_i} and {Q_i}. This approach bypasses the use of the Luneburg's mapping functions and is based only on the postulate that VS is a Riemannian space.

3.2.1. *Multidimensional Mapping according to RiemannianMetric*

Multidimensional scaling (MDS) is a method to give a spatial representation to a set of objects {O_i}, $i = 1, 2, \ldots, N$, on the basis of dissimilarity between O_i and O_j (*e.g.*, Schiffman, *et al.*, 1981). Objects and dissimilarity can be colors and color difference or adjectives and reciprocal of similarity in their meanings, *etc.* Suppose that numerical values d_{ij} (= d_{ji}) are given that represent the degree of dissimilarities between O_i and O_j. From the data matrix $\mathbf{D} = (d_{ij})$, $N \times N$, a configuration of points {P_i} can be constructed in an m-dimensional metric space. Denote by $\hat{d}_{ij}$ interpoint distances between P_i and P_j in this space. Usually, the space is either Euclidean or a space having Minkowski's power metric. The dimension m of the space and {P_i} are determined so that $\hat{d}_{ij}$ reproduces data d_{ij} with a sufficient degree of accuracy. The relationship between d and $\hat{d}$ is assumed to be proportional or monotonically increasing or of some other form. Of course, the proportionality cannot be assumed unless $\mathbf{D}$ satisfies the condition that $d_{ii} = 0$. According to MDS program, the matrix $\mathbf{D}$ can be incomplete in the sense that some elements d_{ij} are left unknown.

Scaling of δ in VS

In this chapter, {O_i} is a visual stimulus pattern {Q_i} in X and dissimilarity is perceived distance δ_{ij} between P_{Vi} and P_{Vj} in the corresponding visual pattern {P_{Vi}} in VS. The S is included as a point

Two Extensions of Luneburg Model 81

Q_0 in $\{Q_i\}$ and hence in $\{Q_i\}$ and $\{P_{Vi}\}$, $i = 0, 1, 2, ..., N$. The S is asked to pay attention to a triplet $\{P_{Vi}, P_{Vj}, P_{V\alpha}\}$ in $\{P_{Vi}\}$. In some triplets, one of P_V is the self. It is not a difficult task for the S to assign a numerical value $r_{i,jk}$ that represents the subjective ratio between two distances from P_{Vi}, δ_{ij} and δ_{ik}, provided that the two distances are not too different. If these ratio assessments are systematically carried out with N points, then we can have scaled values d_{ij}, d_{ik}, d_{0i} etc. such that their ratios reproduce $r_{i,jk}$, etc. The same assessments $r_{i,j\beta}$, $r_{i.\alpha\beta}$, are made with triplets $\{P_{Vi}, P_{Vj} P_{V\beta}\}$, $\{P_{Vi}, P_{V\alpha} P_{V\beta}\}$.

$$r_{i,j\alpha} = \frac{\delta_{ij}}{\delta_{i\alpha}}, \quad r_{i.j\beta} = \frac{\delta_{ij}}{\delta_{i\beta}}, \quad r_{i.\alpha\beta} = \frac{\delta_{i\alpha}}{\delta_{i\beta}} \qquad (3.2.1)$$

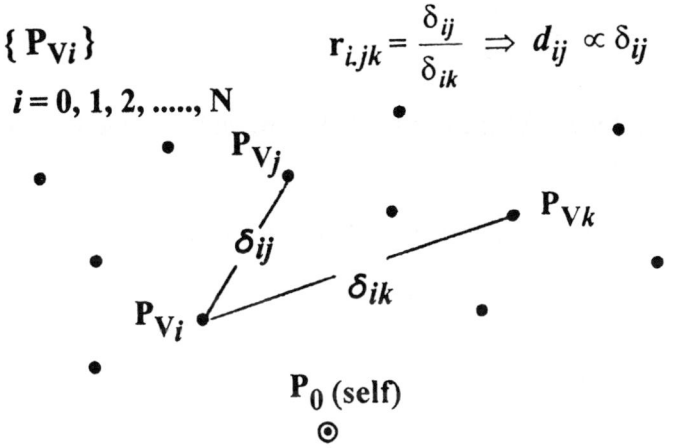

Fig.3.6 Ratio assessment of perceived distance δ

Notice that all δ's are distances from the same points P_{Vi}. Usually the S assigns the numerical value with confidence if r is not too close to 0 or not too far from 5, for instance. It is easy to check whether the assessments are consistently made or not. From the above equations,

$$r_{i,j\alpha} \times r_{i.\alpha\beta} = r_{i,j\beta} \qquad (3.2.2)$$

must hold. As will be shown soon, insofar as visual distances δ's are concerned, the ratio assessments well satisfy this condition. In (3.2.1), δ_{ij} is compared with $\delta_{i\alpha}$ in $r_{i,j\alpha}$ and with $\delta_{i\beta}$ in $r_{i,j\beta}$. Suppose that δ_{ij} is assessed against n different $\delta_{i\alpha}$, and n values of $r_{i,j\alpha}$ are obtained, $\alpha = 1, 2, \ldots, n$. Define the mean value as $x_{ij(i)}$.

$$x_{ij(i)} = \frac{1}{n}\sum_{\alpha} r_{i.j\alpha} = \delta_{ij}\, e_i, \qquad e_i = \frac{1}{n}\sum_{\alpha}\frac{1}{\delta_{i\alpha}} \qquad (3.2.3)$$

Suppose that $x_{ij(i)}$ are obtained for all $i, j = 0, 1, 2, \ldots, N$. By definition, $x_{ii(i)} = 0$. Then, we have a matrix $\mathbf{X} = (x_{ij(i)})$, $(N+1) \times (N+1)$. Elements of its i-th row represent δ_{ij} with the unit e_i, and e_i is different from row to row. Let us denote δ_{ij} by the unit of k-th row.

$$x_{ij(k)} = \delta_{ij}\, e_k = x_{ij(i)}\frac{e_k}{e_i}$$

Take $x_{ki(k)}$, the i-th element in the k-the row and $x_{ik(i)}$, the k-th element in the i-the row,

$$x_{ki(k)} = \delta_{ki}\, e_k, \quad x_{ik(i)} = \delta_{ik}\, e_i, \quad \text{and} \quad \delta_{ki} = \delta_{ik},$$

Hence,

$$\frac{e_k}{e_i} = \frac{x_{ki(k)}}{x_{ik(i)}}$$

and $x_{ij(i)}$ can be converted to $x_{ij(k)}$ by the use of two elements in $\mathbf{X}$.

$$x_{ij(k)} = x_{ij(i)}\frac{x_{ki(k)}}{x_{ik(i)}}$$

When $\mathbf{X}$ is complete and values of all $x_{ij(i)}$ are given, we can define the mean of $x_{ij(k)}$ as d_{ij} that is proportional to δ_{ij} with an arbitrary but common coefficient $\bar{e}$.

$$d_{ij} = \frac{1}{N'}\sum_k x_{ij(k)} = x_{ij(i)}\left[\frac{1}{N'}\sum_k \frac{x_{ki(k)}}{x_{ik(i)}}\right] = e_i\delta_{ij}\left[\frac{1}{N'}\sum_i \frac{e_k\delta_{ki}}{e_i\delta_{ik}}\right]$$
$$= \delta_{ij}\left(\frac{1}{N'}\sum_k e_k\right) = \delta_{ij}\bar{e}, \quad N' = N+1, \quad k = 0, 1, \ldots, N \tag{3.2.4}$$

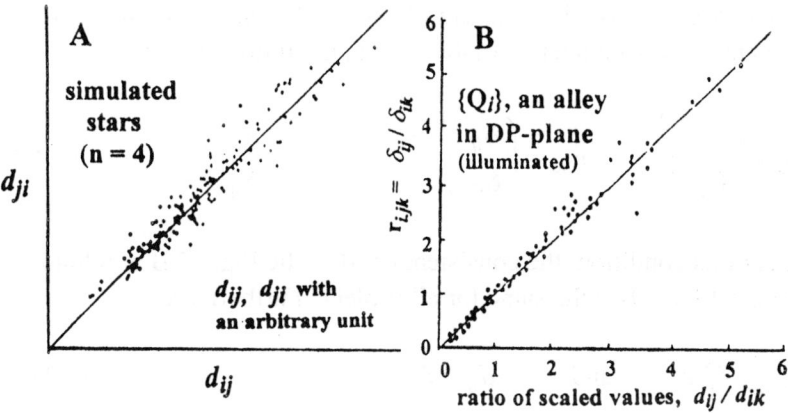

Fig.3.7 Two tests of consistency of ratio assessments of δ

In this way, values of N(N+1) off diagonal elements of **D** = (d_{ij}) are determined, and diagonal elements $d_{ii} = 0$. Because data $x_{ij\,(i)}$ are fallible, d_{ij} and d_{ji} may not be identical. Fig.3.7A shows an example of plotting of d_{ji} against d_{ij} of a S. These are scaled values of perceived distances δ_{ij} in a configuration $\{P_{Vi}\}$ of simulated stars in a dark room. Small light points Q_i, $i = 1, 2, \ldots, 17$, were randomly hung from the ceiling and $\{P_{Vi}\}$ looks like stars in the night sky. As the S was counted as a point Q_0, $N' = 18$ and n in Eq.(3.2.3) was 4 (Indow, 1968). Compared with alley configuration $\{Q_i\}$, this visual pattern is more difficult to make distance assessments. The assessments were made by a large number of S's. In 24 individual cases, plotted points (d_{ij} and d_{ji}) were always scattered along a straight line passing through the origin. Since d_{ij} and d_{ji} are given in terms of an arbitrary unit $\bar{e}$, no scales are shown in Fig.3.7A. This example shows a medium degree of scatter of points. Including assessments with $\{Q_i\}$ of different patterns, such as $\{Q_i\}$ in alley experiments, discrepancy between (d_{ij} and d_{ji}) is always of this degree and the mean of d_{ij} and d_{ji} is re-defined as d_{ij}. As to the scaling procedure when **X** is incomplete, consult Indow and Ida (1975).

Fig.3.7B shows an example of the coincidence between the data $r_{i,jk}$ and ratios of scaled values d_{ij} over d_{ik} in $\{Q_i\}$ of a S, where $\{Q_i\}$ is an alley setting constructed by this S. In general, ratios of scaled values d's reproduce the original data r's in this way, which means that the consistency test (3.2.2) is satisfied.

Suppose the case that, instead of Eq.(3.2.1), ratio assessments $r_{i,j\alpha}$ etc. are generated from latent variables δ's in the following way.

$$r_{i.j\alpha} = \left(\frac{\delta_{ij}}{\delta_{i\alpha}}\right)^{\beta}, \quad r_{i.j\beta} = \left(\frac{\delta_{ij}}{\delta_{i\beta}}\right)^{\beta}, \quad r_{i.\alpha\beta} = \left(\frac{\delta_{i\alpha}}{\delta_{i\beta}}\right)^{\beta} \quad (3.2.5)$$

Under this condition, the consistency test in the Fig.3.7 B will hold if the exponent β (> 0) is the same for all triplets. Furthermore,

$$x_{ij(i)} = \delta_{ij}^{\beta} e_i \quad \text{and} \quad d_{ij} = \delta_{ij}^{\beta} \bar{e} \quad (3.2.6)$$

and the test of consistency in the form of Fig.3.7A will be also passed. Hence, we cannot distinguish, from the tests of consistency in ratio assessments alone, whether $\beta = 1$ or not. However, the following results strongly suggest that β is close to unity.

Additivity of d's for collinear points

Consider three points, P_{V1}, P_{V2}, P_{V3}, that are collinear in this order in VS, which means that if the subject visualizes a line connecting P_{V1} and P_{V3}, the line appears to pass through P_{V2}. It is irrelevant how the line is oriented in VS. The impression of collinearity is direct and unambiguous. If P_{V2} is moved a little off from the line, that impression would be immediately destroyed. The subject may see three lengths, δ_{12}, δ_{23} in addition to δ_{13}, and may feel that δ_{12} and δ_{23} are being concatenated to produce δ_{13}. This fact was counted as one of characteristics of VS (Sec. 2.2.1). However, it is yet to be tested whether scaled values d_{12}, d_{23}, and d_{13} are interlocked in such a way that $d_{13} = d_{12} + d_{23}$. Now, we can test this requirement because, in the experiments reported earlier, we have series of Q's that appear as a perceptually straight line to the S.

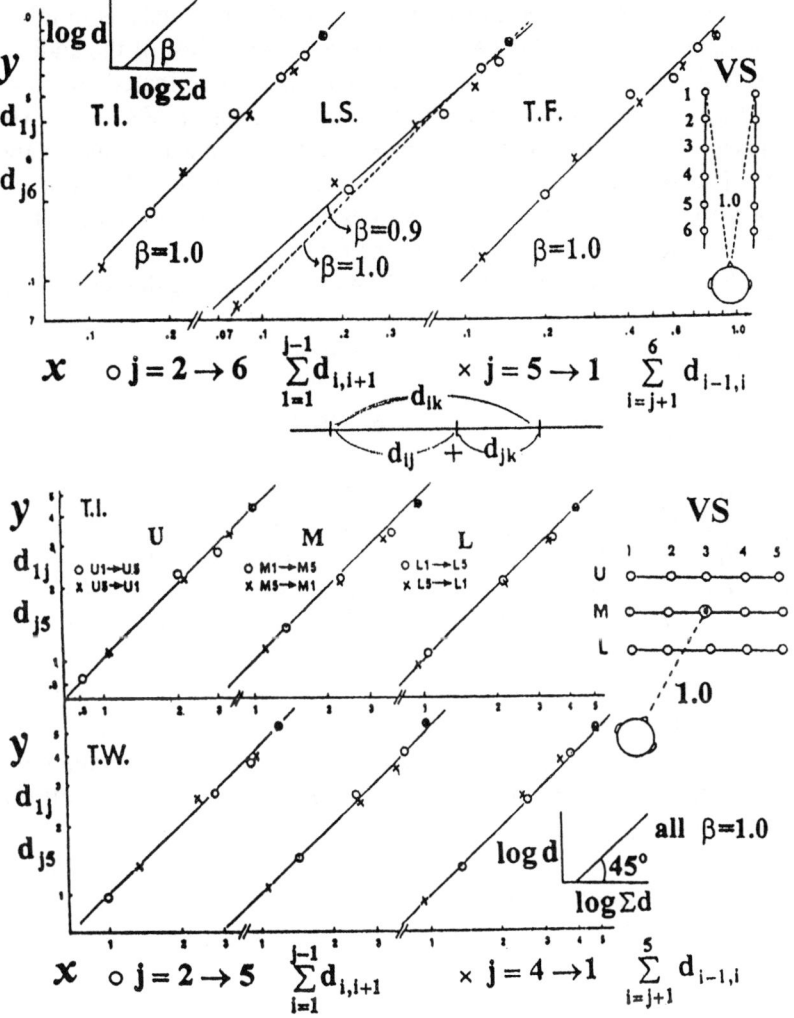

Fig.3.8 Additivity of d's for collinear points

86 Global Structure of Visual Space

A parallel alley configuration $\{Q_i\}_P$ includes series of Q's that appear collinear to the S. Experiments were performed to obtain d_{ij} and d_{0i} from a S with $\{Q_i\}_P$ constructed by that S. The entire set $\{Q_i\}_P$ was presented and remained visible, and various triplets, $\{Q_i, Q_j, Q_\alpha\}$ etc., were indicated to the S. The subject made the assessment $r_{i,j\alpha}$ etc. in Eq. (3.2.1). In this way, (d_{ij}), a complete matrix, was obtained with $\{Q_i\}_P$ and d_{ij} of perceptually collinear Q's were used in the analysis to follow.

The upper figure of Fig.3.8 shows the results of a test of the additivity (three S's) with $\{Q_i\}_P$ in the horizontal DP under illuminated conditions. Each $\{Q_i\}_P$ was constructed by the S so as to appear in VS^2 as shown in the inset. What was shown by black points in Fig.2.1 is one of these $\{Q_i\}_P$. The scaled values d's are normalized so that $d_{01} = 1$, and d_{12} etc. denotes the means of two scaled values of perceived distance between (1, 2) etc. one on the left and one on the right.

Circles and crosses indicate the following plots on log-log coordinates

circles   crosses

y vs. x   y vs. x

d_{12} d_{12} d_{65} d_{65}
d_{13} $d_{12} + d_{23}$ d_{64} $d_{65} + d_{54}$
d_{14} $d_{12} + d_{23} + d_{34}$ d_{63} $d_{65} + d_{54} + d_{43}$
d_{15} $d_{12} + d_{23} + d_{34} + d_{45}$ d_{62} $d_{65} + d_{54} + d_{43} + d_{32}$
d_{16} $d_{12} + d_{23} + d_{34} + d_{45} + d_{56}$ d_{61} $d_{65} + d_{54} + d_{43} + d_{32} + d_{21}$

If $y = x^\beta$ then $\log y = \beta \log x$ and plotting $\log y$ against $\log x$ should be linear with the slope of β. Three lines are best-fitted regression lines of $\log y$ on $\log x$, and the slope β is automatically 1.0 in two S's and 0.9 for one S (L.S.). However, the series of plotted points of L.S. are well fitted by the dotted line of $\beta = 1.0$ also.

The lower figure is with regard to three horizontally parallel lines, U, M, L, on the frontoparallel HP(M3) of the experiment described in Sec. 3.1.2 (two S's). The plotting should be read in the same way and, $\beta = 1.0$ in all the cases.

Insofar as d's under the issue are concerned, clearly the additivity holds and, e.g., $d_{14} = d_{12} + d_{23} + d_{34}$. This would not be possible unless the two conditions hold. One is that $\delta_{ik} = \delta_{ij} + \delta_{jk}$ for collinear points in

VS and the other is that d_{ij} is proportional to δ_{ij} ($\beta = 1.0$ in Eq.3.2.5). It is questionable, however, whether we can still have d of this property for δ beyond the range of VS under discussion.

Construction of $\{P_i\}$ in EM

As to $\{Q_i\}$ of the alley experiment, on the horizontal DP (*e.g.*, Fig.2.1) or on an HP (*e.g.*, Fig. 3.5), the data matrix $\mathbf{D} = (d_{ij})$ was obtained by the procedure described before. It is a part of $\mathbf{D}$ that was used to demonstrate the additivity of d. Now, the entire set of $\mathbf{D}$ is used to try the following approach to VS. In constructing $\{P_i\}$ in EM from the matrix $\mathbf{D}$, $i, j = 0$, 1, 2, ..., N, where the S is included as point Q_0, the assumption of VS being a Riemannian space of constant K under a fixed context is preserved, but the use of Luneburg's mapping functions is bypassed. Two procedures were tried.

MDS into EM (EMMDS)

This is a direct extension of the metric MDS (Torgerson, 1952) to Riemannian metric under the assumption that d is proportional to δ ($\beta = 1.0$ in Eq.3.2.6), $d = u\delta$, $u > 0$. The procedure consists of the following steps and is applicable only when $\mathbf{D}$ is complete and $d_{ii} = 0$.

1. Denote by ρ_{ij} and ρ_0 Euclidean distances between (P_i, P_j) and (P_0, P) in a configuration $\{P_i\}$ in EM. From Eqs. (2.2.1) and (2.2.3),

$$K < 0 \quad q\rho_{ij} = \sqrt{(1 - \tanh^2(cd_{0i}))(1 - \tanh^2(cd_{0j}))} \sinh(cd_{ij})$$

$$q\rho_0 = \tanh(cd_0) \quad (3.2.7a)$$

$$c = \frac{q}{u} = \frac{\sqrt{-K}}{2u}$$

$$K = 0 \quad q\rho_{ij} = d_{ij}, \quad q\rho_0 = d_0 \quad (3.2.7b)$$

$$K > 0 \quad q\rho_{ij} = \sqrt{(1 + \tan^2(cd_{0i}))(1 + \tan^2(cd_{0j}))} \sin(cd_{ij})$$

$$q\rho_0 = \tan(cd_0) \tag{3.2.7c}$$

$$c = \frac{q}{u} = \frac{\sqrt{K}}{2u}$$

Specify the sign of K and a value of c unless K = 0. Then, a matrix $(q\rho_{ij})$ under the given sign of K can be defined from **D**, where $i, j = 0, 1, 2, \ldots, N$.

2. Consider N + 1 vectors to P_i from the centroid of $\{P_i\}$ and denote its coordinates as $a_{i\alpha}$, $\alpha = 1, 2, \ldots, m$. Denote by b_{ij} the scalar products of vectors from the centroid to P_i and P_j.

$$b_{ij} = \sum_\alpha a_{i\alpha} a_{j\alpha}$$

Denote $\mathbf{B} = (b_{ij})$, (N+1)×(N+1), and $\mathbf{A} = (a_{i\alpha})$, (N+1) × m. Then,

$$\mathbf{B} = \mathbf{A}\mathbf{A}^T \tag{3.2.8}$$

where $\mathbf{A}^T$ means the transpose of **A**. The left side matrix **B** can be obtained from $(q\rho_{ij})$

$$b_{ij} = \frac{1}{2} \left(\frac{1}{N+1} \sum_i (q\rho_{ij})^2 + \frac{1}{N+1} \sum_j (q\rho_{ij})^2 - (q\rho_{ij})^2 - \frac{1}{(N+1)^2} \sum_i \sum_j (q\rho_{ij})^2 \right)$$

3. The matrix **A** in the right side of (3.2.8) is obtained by taking eigenvalues λ_α and eigenvectors $\mathbf{v}_\alpha = (v_{0\alpha}, v_{1\alpha}, \ldots, v_{N\alpha})^T$ of **B**, $\alpha = 1, 2, \ldots, N$ at most. Because **B** is symmetric, all λ_α are real values. If $\lambda_1 > \lambda_2 > \ldots > \lambda_m$ are significantly positive and the rest can discarded as 0, $\{P_i\}$ with interpoint distances $q\rho_{ij}$ can be defined in EM^m around the centroid. Eigenvectors $\mathbf{v}_\alpha$ are given with arbitrary units. Suppose each is normalized such that $\lambda_\alpha = \sum_i v_{i\alpha}^2$. Then, the least square solution of **A** that satisfies (3.2.8) is

$$a_{i\alpha} = \frac{\sqrt{\lambda_\alpha}}{\sqrt{\sum_i v_{i\alpha}^2}} v_{i\alpha}, \quad \alpha = 1, 2, \ldots, m$$

That is to say, if the distance between P_i and P_j is denoted as

$$q\hat{\rho}_{ij} = \sqrt{\sum_\alpha (a_{i\alpha} - a_{j\alpha})^2} \qquad (3.2.9)$$

then, RMS of $(q\rho_{ij} - q\hat{\rho}_{ij})$ is the smallest under the given value of c.

4. By changing the value of c, repeat the processes from 1 to 3 until a satisfactory coincidence between $q\rho_{ij}$ and $q\hat{\rho}_{ij}$ is obtained. Since the size of $\{P_i\}$ changes according to the value of c, the criterion is [RMS/size of $\{P_i\}$]. Usually the sign of K is anticipated. If necessary, however, try the different sign of K or K = 0. The analysis of **D** by this procedure with $\{Q_i\}$ consisting P-and D-alleys and H-curves in the horizontal DP (three S's) is given in Indow (1982). In each S, K < 0, m = 2 and [RMS/size of $\{P_i\}$] became minimum and average of $(q\rho_{ij} - \hat{d}_{ij})$ became close to 0 at the same value of c.

Once the final solution **A** is obtained, we can plot $\{P_i\}$ according to $(a_{i\alpha})$ with the origin at the centroid. If necessary, we can shift the origin from the centroid to P_0 and change these arbitrary coordinates (a_1, a_2, a_3)

to (ξ, η, ζ). The theoretical curves in EM can be fitted to $\{P_i\}$. Furthermore, if necessary, the mapping functions under the given condition can be determined by comparing $\{P_i\}$ in EM and $\{Q_i\}$ in X. Since u and q are not separable in Eqs.(3.2.7a, c), we cannot define K with the unit given in Eq.(2.2.4) unless we have the estimated value of max $q\rho_0$.

Direct mapping through Riemannian powered distance (DMRPD)

The EMMDS gives us a $\{P_i\}$ in EM in which interpoint distances $q\hat{\rho}_{ij}$ (Eq.3.2.9) numerically reproduce $q\rho_{ij}$ that are defined from data d_{ij} with a value of c. According to Eqs. (2.2.1) to (2.2.3), lengths of chord $q\hat{\rho}_{ij}$ between P_i and P_j are supposed to represent δ_{ij}, interpoint distances in the perceived configuration $\{P_{Vi}\}$. However, the perceived geometrical structure of $\{P_{Vi}\}$ is not explicitly included in the criterion to define $\{P_i\}$. The second procedure, DMRPD, starts from an initial configuration $\{P_i\}^{(0)}$ that is anticipated to represent $\{P_{Vi}\}$ in EM and adjusts its form so that its interpoint distances $\hat{d}_{ij}$ closely reproduce data d_{ij}. The S must be included as P_0 in $\{P_i\}$. In this procedure, the data matrix **D**, (N + 1) × (N + 1), does not need to be complete. The initial configuration, $\{P_i\}^{(0)}$ can be given in terms of (ξ, η, ζ) with an arbitrary unit. The unit of $\{\xi_i, \eta_i, \zeta_i\}$ is not required to satisfy the condition that the radius of the basic circle is $1/q$ (Fig.3.2), because Eq.(3.1.2) is independent of this definition of the unit. As in EMMDS, in order to denote the position of P_i, instead of $\{\xi_i, \eta_i, \zeta_i\}$, let us use Cartesian coordinates $a_{i\alpha}$, $i = 0, 1, 2, ..., N$, $\alpha = 1, 2$ or $1, 2, 3$.

According to the nature of $\{P_{Vi}\}$, the value of m is anticipated. When $\{Q_i\}$ is on an HP(M3) (an example in Fig.3.10), m = 3. When $\{Q_i\}$ is on the horizontal DP (an example in Fig.3.9), m = 2. From the coordinate matrix $(a_{i\alpha})$, ρ_{ij}, ρ_{0i}, are determined and once a value of q is taken, δ_{ij}, δ_{0i}, are defined by Eqs.(2.2.1) – (2.2.3). Denote δ_{ij}, and δ_{0i} as $\hat{d}_{ij}$ and $\hat{d}_{0i}$ that should represent d_{ij} and d_{0i} in **D**. Then, we can plot d_{ij} against $\hat{d}_{ij}$. It is natural to take $\hat{d}$ on the abscissa and d on the ordinate. However, the independent variable is d because these are fixed as the data. To be adjusted are $\hat{d}_{ij}$ (= δ_{ij}) as functions of ($a_{i\alpha}$) and q. Instead of assuming that d is proportional to δ, a more general approach is taken and a curve

$$\tilde{d} = A d^B \qquad (3.2.10)$$

is used to represent the trend in the scatter diagram, d_{ij} vs. $\hat{d}_{ij}$. Then, the program tries to minimize the criterion often used in nonmetric MDS

$$\text{Stress} = \sqrt{S} = \sqrt{\frac{P}{Q}}, \quad P = \sum_{i>j}(\hat{d}_{ij} - \tilde{d}_{ij})^2, \quad Q = \sum_{i>j}\hat{d}_{ij}^2 \qquad (3.2.11)$$

by adjusting values of a_{ij}, q, and A, B in (3.2.10). The optimization of these values is carried out by the method of steepest descent.

Initial values of unknowns, $q^{(0)}$ (under the given sign of K) and an initial configuration $(a_{i\alpha})^{(0)}$ are specified under the given dimensionality m. From these values, $\hat{d}$'s are calculated. By plotting d's against $\hat{d}$'s and fitting Eq.(3.2.10), we have $\tilde{d}$'s as well as initial values of the two parameters, $A^{(0)}$ and $B^{(0)}$. Thus, $\text{Stress}^{(0)}$, the first value of Stress, is determined. Then, $q^{(0)}$, $(a_{i\alpha})^{(0)}$, $A^{(0)}$ and $B^{(0)}$ are adjusted to the second set of values, $q^{(1)}$, $(a_{i\alpha})^{(1)}$, $A^{(1)}$ and $B^{(1)}$, so as to make $(\text{Stress}^{(0)} - \text{Stress}^{(1)})$ largest. These iterations I = 1, 2, ... are continued until the decrease from $\text{Stress}^{(I-1)}$ to $\text{Stress}^{(I)}$ becomes negligible, and the last set of adjusted values are taken as q, $(a_{i\alpha})$, A and B. The outline of this program is given at the end of Sec.3.2.3. When the scatter of points around $\tilde{d}$ is small enough, $\tilde{d}$ can be written as a function of $\hat{d}$.

$$\tilde{d} = \alpha \hat{d}^\beta, \quad \alpha = A^{\frac{-1}{B}}, \quad \beta = \frac{1}{B} \qquad (3.2.12)$$

3.2.2. Experimental Results

P- and D- alleys and H-curves in the horizontal DP

Fig.2.1A shows $\{Q_i\}$ that includes all Q_i on P-and D-alleys and on 5 H-curves in the DP(θ=0) constructed in the dark by an inexperienced subject (T.F). After this experiment was completed, this S participated in the second experiment to yield **D** with this $\{Q_i\}$. The left plot in Fig.3.9 is $\{P_i\}$ that was obtained from **D** by the EMMDS. In this experiment, a particularly complicated procedure was employed to have $\{P_i\}$ for the following reason.

The configuration $\{Q_i\}$, Fig.2.1A, consists of 28 Q's including the S as Q_0. The procedure EMMDS requires **D** to be complete, which means (28 × 27)/2 = 378 δ's must be scaled as d's. If all Q's are simultaneously presented, the appearance of $\{Q_i\}$ is too confusing to the S, and some pairs, e.g., black circle and white diamond, on each side of H_A-curve, are too close together. Hence, the following procedure was tried in this experiment.

Out of $\{Q_i\}$, i = 0, 1, 2, ..., 27, five subsets $\{Q_i\}_S$, each consisting of 12 Q's, were selected, i = 0, 1, 2, ...,12, and s = 1, 2, ..., 5. Three points, Q_1, Q_2, and Q_0 (the S), were included in each set and remaining Q's were distributed so that each Q_i was contained in two different subsets at least. At a time, one set $\{Q_i\}_S$ was presented and a complete matrix $\mathbf{D}_S = (d_{ij})$ was obtained (n = 2 in Eq.3.2.3). This is a time-consuming experiment and it took about 10 hours to obtain $\mathbf{D}_S$, s = 1, 2, ..., 5, from a S. In each set, d_{01} was defined as unity. In EMMDS, what is to be optimized is the value of c only and its optimum value was determined so that $q\rho_{ij}$ defined from d_{ij} by Eq.(3.2.5) numerically coincide most closely with Euclidean interpoint distances $q\hat{\rho}_{ij}$ of $\{P_i\}$'s in all the five sets. From each $\mathbf{D}_S$, a configuration $\{P_i\}_S$, i = 0, 1, 2, ..., 12, was obtained in EM^2 with the use of a value of c. It was found that the two criteria stated in Step 4 of EMMDS were satisfied by the same value of c for all the five sets. Then, $\{P_i\}_S$, s = 1, 2, ..., 5, were synthesized by the use of several common points as pivots to give one $\{P_i\}$ for the whole sets where i = 0, 1, 2, ..., 27. The left upper figure in Fig.3.9 is this configuration $\{P_i\}$. The lower plot shows an example of the coincidence between interpoint distances $q\hat{\rho}_{ij}$ (3.2.9) and $q\rho_{ij}$ obtained from the data d_{ij} and c (3.2.7a) in a $\{P_i\}_S$. The correlation coefficient r between $q\rho_{ij}$ and $q\hat{\rho}_{ij}$ is 0.98 throughout the five sets. In the upper figure (the coordinate axes, qξ and qη), black and white symbols

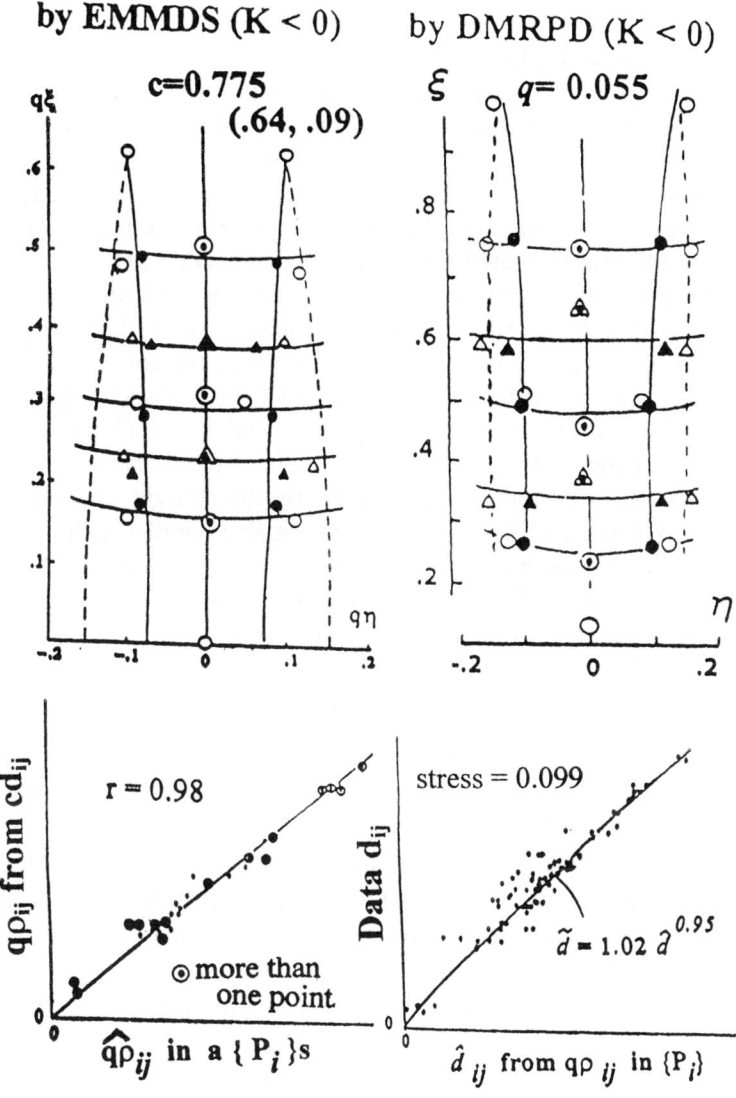

Fig.3.9 P-and D-alleys and H-curves in EM^2 that were obtained without using the Luneburg's mapping functions

respectively show $\{P_i\}_P$ and $\{P_i\}_D$, and symbols of the same figure shows $\{P_i\}_{H\alpha}$. Schematic illustrations of the respective theoretical curves are shown. If the fit is perfect, black symbols should be at the intersections of P-alley and H_α-curves, whereas white symbols should be at the intersections of D-alley and H_α-curves.

A disappointing feature of this $\{P_i\}$ is that two groups of symbols, black and white, were not well separated despite the fact that $\{Q_i\}_P$ and $\{Q_i\}_D$ are clearly separated in X^2. The situation was the same with other two S's (Indow, 1982). This may be due to either inadequacy of the procedure of constructing $\{P_i\}$ from five $\{P_i\}_S$ or the logic of EMMDS. This is the reason that the second procedure DMRPD was developed. Before going to the analysis of the same data by DMRPD, the right side plots, it would be in order to add two comments on the left side plots.

The result presented on the left side is under the constraint that m = 2 and K < 0 (Eq.3.2.7a). In Fig.13 of Indow (1982) is shown $\{P_i\}$ that was obtained under the constraint that m = 2 and K = 0 (Eq.3.2.7b). If the perceived pattern $\{P_{Vi}\}$ under discussion is of Euclidean structure, P-and D- alleys must be represented in EM^2 by the straight line parallel with the $q\xi$-axis and each H-curve is a straight line parallel with the $q\eta$-axis. Clearly, it was not the case in all the three S's. Namely, $\{P_{Vi}\}$ that the S was perceiving while constructing P-and D-alleys and also assessing distances is far from Euclidean structure.

In Fig.3.9, the optimized value of c was 0.775. According to Eq.(3.2.7a), $c = \sqrt{-K}/2u$. However, there is no way to have the value of u and we cannot define K from c. From Fig.2.4, max $q\rho_0 = \sqrt{-K}$. This is the boundary of VS under the given condition at which $q\rho_0$ ceases to increase as a function of distance $e(Q_0, Q)$ in X or for which the convergence angle γ for Q becomes 0. Hence, if max $q\rho_0$ is located in Fig.3.9, we can have K. Though we can plot $q\rho_{0i}$ against $e(Q_0, Q_i)$ or γ_i, an extensive extrapolation is necessary to locate max $q\rho_0$ (see Fig.15 in Indow, 1982). The best estimate in this case was max $q\rho_0 = 0.90$ and hence K = -0.81. We have no way to compare this value of K with K = -0.37 given in Fig.2.1, because the latter is meaningful only with σ = 21.86. The present analysis is trying to bypass the Luneburg's mapping functions. That K by the present procedure was closer to -1.0 than K with σ was found with the other two S's too. In order to fit theoretical equations in EM, it is not necessary to dissolve c in to K and u.

The right upper figure in Fig.3.9 show $\{P_i\}$ obtained by DMRPD from $\mathbf{D} = (d_{ij})$, $i, j = 0, 1, 2, \ldots, 27$, where d_{ij} are collected from $\mathbf{D}_S$, $s = 1, 2, \ldots, 5$ in the EMMDS analysis stated above. All d's for the same (Q_i and Q_j) in separate $\mathbf{D}_S$ were averaged to give d_{ij}. This $\mathbf{D}$ is far from being complete. Out of $378 = (28 \times 27)/2$ off-diagonal elements, only 130 values were available, about a third of the total number. This is not an obstacle in DMRPD. The upper right figure is $\{P_i\}$ based on the final $(a_{i\alpha})$ after 11 iterations, $i = 0, 1, 2, \ldots, 27$ and $\alpha = 1, 2$. The symbols and curves should be read in the same way as in the left plot. The coordinates ξ and η are the first and second axes in $(a_{i\alpha})$ with an arbitrary unit. The initial configuration $\{P_i\}^{(0)}$ and the initial value $q^{(0)}$ were defined in reference to (a_{ij}/c) by EMMDS, and $B^{(0)} = 1$. It was found that the program is not very powerful to appropriately adjust the value of q during the iterations. Hence, it is recommendable to repeat the analysis with various initial values of $q^{(0)}$ to find the most satisfactory result. In this case, $q = 0.055$ and through 11 iterations, Stress$^{(0)} = 0.252$ was reduced to Stress $^{(11)} = 0.099$ under the constraint that $K < 0$. In the lower plot, data d_{ij} are plotted against $\hat{d}_{ij}$ that were obtained from $\{P_i\}^{(11)}$. Since d is defined with an arbitrary unit, the scale is not given in the plot. Because $B^{(11)} = 1.053$ in Eq.(3.2.10), $\beta = 0.950$ in Eq.(3.2.12) and the curve $\tilde{d}$ is very slightly convex upward.

Ph-alley on a frontoparallel plane HP

As an example, horizontal alleys $\{Q_i\}$ on a frontoparallel plane HP(M3) of S.W. was given in Fig.3.5. Two other S's, T.I. and T.W., who are more experienced than S.W., participated in this experiment. The results of the three S's are in agreement in the sense that $\{Q_i\}_P$ and $\{Q_i\}_D$ coincide with each other perfectly (Fig.1 in Indow and Watanabe, 1988). However, forms of three horizontal series in X^3 are not the same. The series are concave toward the S in S.W. and T.W. whereas these are convex in T.I. This is due to individual differences in the inflection point at which H-curve change from concavity to convexity on the horizontal DP (see Fig.2.1). The coincidence between $\{Q_i\}_P$ and $\{Q_i\}_D$ implies that the frontoparallel plane is structured according to Euclidean geometry. Now, this hypothesis is tested with $\{P_{Vi}\}$ generated from $\{Q_i\}$ by checking whether perceived distances δ's behave as Euclidean distances. The two S's, T.I. and T.W., assessed δ's with $\{Q_i\}$ that had been constructed by themselves.

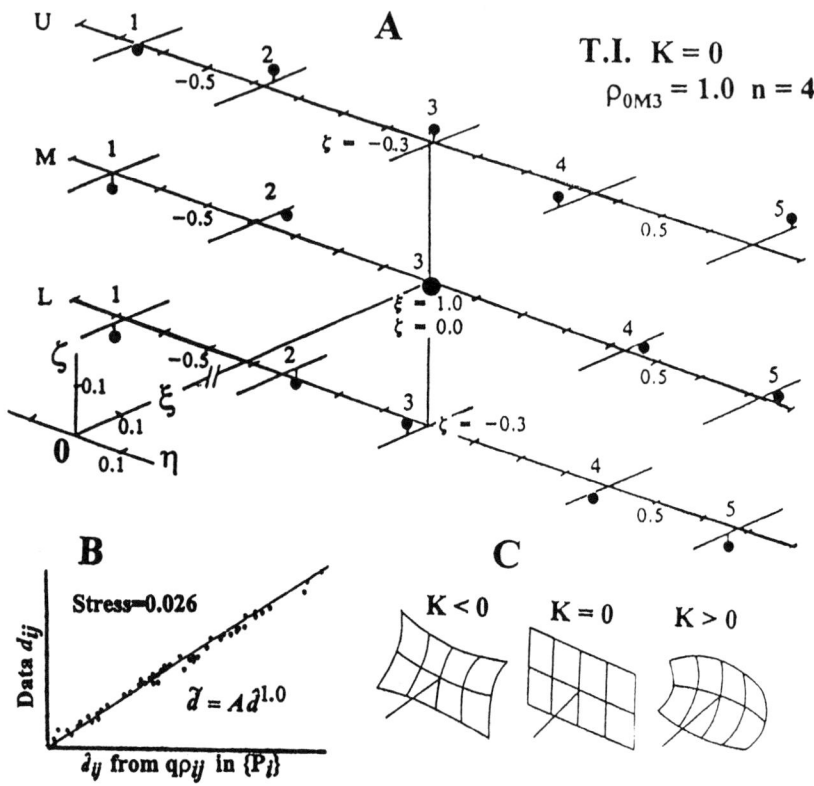

Fig.3.10 $\{P_i\}$ of Ph-alley in EM^3 by DMRPD

Since Q's for Ph- and Dh-alleys were so close, only $\{Q_i\}_P$ for Ph-alleys (15 Q's plus the S) was presented, and the complete **D** was obtained by the procedure of ratio assessments. All off-diagonal elements d_{ij} in **D**, $(16 \times 15)/2 = 120$, were based on four repeated assessments (n = 4). The optimization was tried under the three constraints, K < 0, K = 0, and K > 0, and with various initial values. The configuration $\{P_i\}$ of T.I. is given in Fig.3.10A that was constructed in EM^3 from **D** by DMRPD.

The smallest value of Stress (0.026) was found under the constraint K = 0 and B = 1.00. As K = 0, $\hat{d}$ is equal to the interpoint distance qρ and it is proportional to data d (Fig.3.10B). The next smallest Stress (0.031) was under the constraint K < 0 (q = 0.45) and B = 0.99. In the case of T.W., almost the same results were obtained under the three constraints. For K < 0, K = 0, K > 0, Stress values were 0.029 (q = 0.36), 0.030, 0.030 (q = 0.40) and B were 0.95, 1.00, 1.00, respectively. The hypothesis of K being 0 is not discarded. The value of Stress is independent of unit of $\hat{d}_{ij}$. In right plot of Fig.3.9, {P_{Vi}} for (P-and D-alleys and H-curves in DP($\vartheta_V = 0$)) the stress was 0.099. The general level of Stress in the present HP was about 0.03. That the reproducibility of d_{ij} by $\hat{d}_{ij}$ is three times better will be apparent if the lower right plot of Fig. 3.9 is compared with Fig.3.10B. Furthermore, in the present case, the value of B is unity and the curve $\tilde{d}$ is a straight line passing through the origin in the plot of d_{ij} against $\hat{d}_{ij}$ (Fig.3.10B).

Fig.3.10C illustrates how the theoretical alleys on the horizontal HP(M3) are represented as {$\hat{P}_i$} in EM^3. Their forms differ according to the sign of K. If K = 0, it is a flat mesh. It is a mesh on a convex or concave surface (toward the S) according to whether K < 0 or K > 0. Either {P_i} obtained by DMRPD from data **D** or {$\hat{P}_i$} determined by K has nothing to do with the pattern of {Q_i} in X^3. In the case of T.I., the conclusion of K being 0 is supported not only by the Stress value but also by the form of {P_i} of Fig.3.10A, because it is closest to the middle pattern in Fig.3.10 C. In the case of T.W., three geometries gave almost the same Stress values and {P_i} was of such a form with which it is difficult to say which pattern in C is closest. In any event, we can have {P_i} under the constraint of K = 0 in which $\hat{d}_{ij}$ reproduce data d_{ij} well.

3.2.3. Concluding Remarks to Sec.3.2

It is true that if we can scale perceptual distances δ in a perceived pattern {P_{Vi}}, it will open a way for flexible approaches to VS. However, many improvements must be made as to the scaling procedure of δ to obtain **D** = (d_{ij}) and also to algorithm of constructing {P_i} from **D**.

All **D**'s in Sec.3.2 were obtained by the procedure described in Sec.3.2.1. With a pattern $\{P_{Vi}\}$ being perceived, $i = 0, 1, 2, \ldots, N$, the S is asked to pay attention, one by one, to a triplet $\{P_{Vi}, P_{Vj}, P_{V\alpha}\}$ to make an assessment on subjective ratio $\delta_{ij}/\delta_{i\alpha}$ (Fig.3.6). The whole configuration $\{P_{Vi}\}$ must remain visible because δ's must be interpoint distances in this pattern. Each triplet is notified to the S by blinking the three when Q's are light points or verbally telling their locations in $\{Q_i\}$, e.g., "P_{Vi} is the second from left in the top row, $P_{V\alpha}$ is the third in the second row, and P_{Vj} is furthest right in the top row". However, because of the dynamic nature of VS (VS5 in Sec.1.1.1), paying attention to a triplet may bring about some effect to the structure of this part in $\{P_{Vi}\}$. In an experiment in which real stars in the night sky were Q's, when the S gazed at a star, sometimes the S felt as if that star retreated into the sky. This experiment will be reported in Sec.4.1.3.

Furthermore, to obtain a sufficient number of *d*'s in this way is a tedious procedure when N is large. This was the reason that, in the left plot of Fig. 3.9, the whole configuration of 28 Q's was divided into five subsets of $\{Q_i\}$'s, each consisting of 12 Q's, and then the results were synthesized to define $\{P_i\}$ of 28 points. It was found that this procedure is not quite appropriate. It is likely that the pattern $\{P_{Vi}\}$ the S saw in each subset did not well represent the entire pattern $\{P_{Vi}\}$. In short, we need an assessment procedure that is more efficient and has smaller distracting effects on the perceptual pattern $\{P_{Vi}\}$ of the entire configuration $\{Q_i\}$.

The reason that the metric MDS was used in EMMDS is that no iterative procedure is necessary if a subprogram to obtain eigenvalues and eigenvectors of **B** (Eq.3.2.8) is available. However, the use of metric MDS restricts the applicability of EMMDS to a complete **D** only. If the metric MDS is replaced by another MDS algorithm, EMMDS can be altered to be applicable to an incomplete data **D** matrix.

The outline of DMRPD is as follows.

$$\mathbf{D} = (d_{jk}), \quad d_{jj} = 0, \quad j < k,$$

$$\text{Stress} = \sqrt{S}, \quad S = \frac{P}{Q}, \quad P = \sum_{j<k}\left(\hat{d}_{jk} - \tilde{d}_{jk}\right)^2 \quad Q = \sum_{j<k}\hat{d}_{jk}^2$$

$$\hat{d}_{jk} = \delta_{jk} \quad \text{from } \rho_{jk}(q, a_{j\alpha}) \text{ by Eqs.(2.2.1) – (2.2.3)}$$

$\tilde{d}_{jk} = A d_{jk}^B$ from the plotting d against $\tilde{d}$

Variables X_g to be optimized are q, $a_{j\alpha}$ ($j = 0, 1, 2, ..., N$, $\alpha = 1, 2, .., m$), A and B.

$$X_g^{(I+1)} = X_g^{(I)} - (\text{appropriate weight}) C \nabla S^{\frac{1}{2}}(X_g), \quad I = 0, 1, 2, ...$$

$$C = [\Sigma_g (\nabla S^{\frac{1}{2}}(X_g))^2]^{-\frac{1}{2}}$$

$$\nabla S^{\frac{1}{2}}(X_g) = \frac{\partial \sqrt{S}}{\partial X_g} = \frac{1}{2\sqrt{S}} \frac{\partial S}{\partial X_g}$$

When $X_g = A$ or B,

$$\frac{\partial S}{\partial X_g} = \frac{\partial}{\partial X_g} \frac{P(X_g)}{Q} = \frac{\partial S}{\partial P} \frac{\partial P(X_g)}{\partial X_g} = \frac{1}{Q} \frac{\partial P(X_g)}{\partial X_g}$$

$$\nabla S^{\frac{1}{2}}(A) = -\frac{2}{Q} \sum_{j<k} (\hat{d}_{jk} - \tilde{d}_{jk}) \frac{\tilde{d}_{jk}}{A}$$

$$\nabla S^{\frac{1}{2}}(B) = -\frac{2}{Q} \sum_{j<k} (\hat{d}_{jk} - \tilde{d}_{jk}) \tilde{d}_{jk} \ln d_{jk}$$

When $X_g = q$ or $a_{j\alpha}$

$$\frac{\partial S}{\partial X_g} = \frac{\partial}{\partial X_g} \frac{P(X_g)}{Q(X_g)} = \frac{\partial S}{\partial P} \frac{\partial P(X_g)}{\partial X_g} + \frac{\partial S}{\partial Q} \frac{Q(X_g)}{\partial X_g} = S \left(\frac{1}{P} \frac{\partial P}{\partial X_g} - \frac{1}{Q} \frac{\partial Q}{\partial X_g} \right)$$

$$\nabla S^{\frac{1}{2}}(q) = S \left[\frac{1}{P} \sum_{j<k} (\tilde{d}_{jk} - \hat{d}_{jk}) Z(q)_{jk} - \frac{1}{Q} \sum_{j<k} \hat{d}_{jk} \right]$$

$$\nabla S^{1/2}(a_{j\alpha}) = S\left[\frac{1}{P}\left\{\sum_{i=1}^{j-1}(\hat{d}_{ij} - \tilde{d}_{ij})Z(a_i)_j + \sum_{k=j+1}^{N}(\hat{d}_{jk} - \tilde{d}_{jk})Z(a_k)_j\right\}\right.$$
$$\left. - \frac{1}{Q}\left\{\sum_{i=1}^{j-1}\hat{d}_{ij}Z(a_i)_j + \sum_{k=j+1}^{N}\hat{d}_{jk}Z(a_k)_j\right\}\right]$$

When $K < 0$,

$$Z(q)_{jk} = \frac{1}{q}\left(\frac{G_j^2 + G_k^2}{(G_j G_k)^2} - 1\right)\tanh \hat{d}_{jk}$$

$$Z(a_i)_j = \left(\frac{-(a_{i\alpha} - a_{j\alpha})}{\rho_{ij}^2} + \frac{q^2 a_{j\alpha}}{G_j^2}\right)\tanh \hat{d}_{jk}$$

$$Z(a_k)_j = \left(\frac{(a_{j\alpha} - a_{k\alpha})}{\rho_{jk}^2} + \frac{q^2 a_{j\alpha}}{G_j^2}\right)\tanh \hat{d}_{jk}$$

When $K = 0$, $\nabla S^{1/2}(q)$ is not involved.

$$Z(a_i)_j = \frac{-(a_{i\alpha} - a_{j\alpha})}{\rho_{ij}}$$

$$Z(a_k)_j = \frac{(a_{j\alpha} - a_{k\alpha})}{\rho_{jk}}$$

When $K > 0$

$$Z(q)_{jk} = \frac{1}{q}\left(\frac{G_j^2 + G_k^2}{(G_j G_k)^2} - 1\right)\tan \hat{d}_{jk}$$

$$Z(a_i)_j = \left(\frac{-(a_{i\alpha} - a_{j\alpha})}{\rho_{ij}^2} - \frac{q^2 a_{j\alpha}}{G_j^2}\right)\tan \hat{d}_{jk}$$

$$Z(a_k)_j = \left(\frac{(a_{j\alpha} - a_{k\alpha})}{\rho_{jk}^2} - \frac{q^2 a_{j\alpha}}{G_j^2}\right)\tan \hat{d}_{jk}$$

4. Visual Space under Natural Conditions

Geometrical patterns dealt with in Chapters 2 and 3 are very different from those we see in VS under natural conditions. Parallel and distance-alleys, $\{Q_i\}_P$ and $\{Q_i\}_P$, were constructed only in a dark or illuminated frameless DP plane extending forwards from the body. I do not doubt that $\{Q_i\}_D$ would lie outside $\{Q_i\}_P$ even if these are constructed in an outdoor field with plenty of spatial cues, provided the subject S understands the requirement correctly and such a case is excluded that regular grid marks are visible on the ground. If $\{Q_i\}$'s are large, these may not be fitted well by the equations based on the Luneburg's mapping functions. However, the fact of $\{Q_i\}_D$ being outside $\{Q_i\}_P$ is interpreted as showing that the perceived patterns $\{P_{Vi}\}_P$ and $\{P_{Vi}\}_D$ on a DP are structured according to hyperbolic geometry. In this chapter, the geometrical properties of perceived patterns in VS under natural conditions will be discussed.

When we are in an open outdoor space, our VS is structured as exemplified in Fig.1.1. The self is localized with regard to the perceived ground and sky. These constitute the largest possible framework in VS. Let us jump from the appearance of $\{Q_i\}$'s in laboratory experiments to the appearance of the sky and ground.

4.1. The Perceived Sky and Ground

In daytime, the proximal stimulus of the sky is the light of the sun scattered by the layer of air around the earth. If there is no air, we will see the sun in the dark space as an astronaut sees the shining earth in the dark space. According to the law of Rayleigh, air particles scatter short wavelength energy 16 times more than long wavelength energy. This is the physical cause for the sky to appear blue. When the physical sky is clear, the proximal stimulus has no texture as that in the Ganzfeld experiment (VS5 in Sec.1.1.1) and the perceived blue sky is of the aperture color mode. We can only tell where the inner surface of sky starts in a given direction and, differing from the surface of a wall, the sky appears to have indefinite depth. We feel as if fingers will penetrate into the sky. Hereafter, the sky means the sky as a percept in VS.

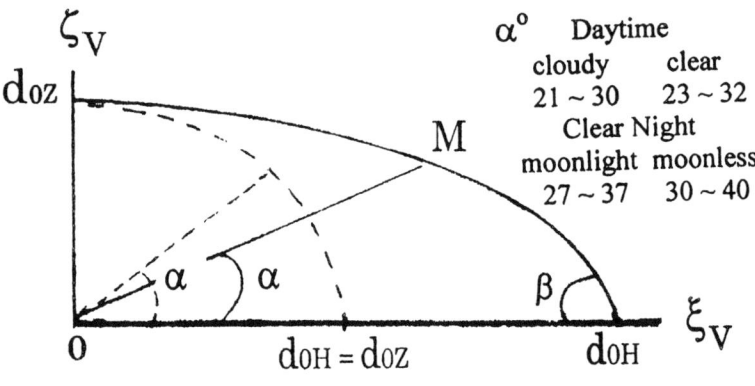

Fig.4.1 Bisection of the sky

4.1.1. Bisection of the Sky

The daytime sky is generally regarded as a bowl flattened in the zenith direction. Fig.4.1 shows its slice in the ξ_V and ζ_V - plane where the sky is perceived to meet the ground or the ocean at the horizon with an angle of β. The attempt to determine the shape of the sky has a long history. According to Filehne (1912), as early as 1728 Smith tried a psychophysical observation to bisect the sky. Denote by d_{0Z} and d_{0H} the distance to the zenith and that to the horizon, and by α the half-arc angle that bisects the "imaginary arc" from the horizon to zenith. If the sky is a quarter circle (dotted curves), then $d_{0Z} = d_{0H}$ and $\alpha = 45°$. Smith found $\alpha = 23°$, which implies that $d_{0Z} < d_{0H}$. Neuberger (1951) summarized in his Table 1 values of the half-arc angle α obtained by four investigators (from 1890 – 1943). These values are further summarized in the right side in Fig.4.1. No doubt, $d_{0Z} < d_{0H}$ in daytime as well as at night, and the night sky is less flattened. Some investigators tried to directly assess the ratio of d_{0H} over d_{0Z}. This assessment is less direct than the ratio assessments discussed in Fig.3.6, because we cannot see both d_{0H} and d_{0Z} in a single glance. This is the reason that most investigators determined the value of α in order to obtain the information on d_{0H} / d_{0Z}.

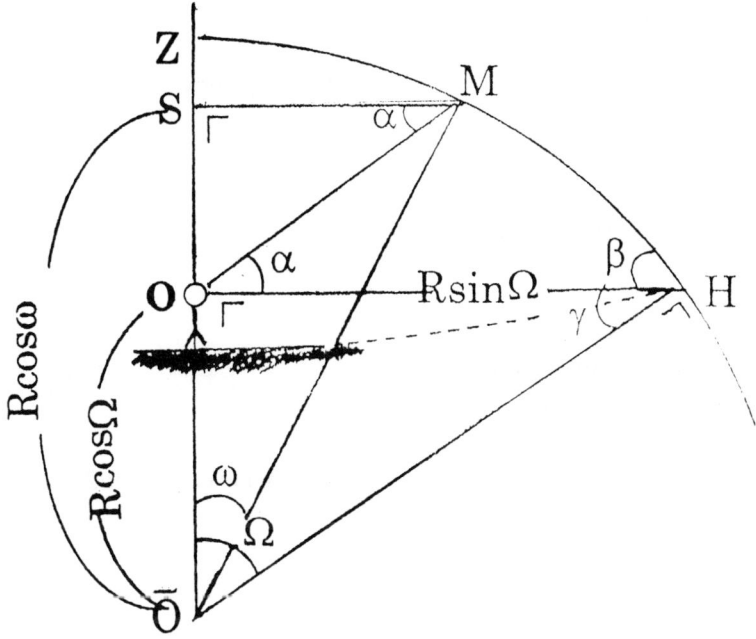

Fig.4.2 The sky as a part of a circle having the center below the ground

If we assume that the sky in Fig.4.1 is a part of a circle of radius R around the center $\overline{O}$, then we can relate the half-arc angle α to the ratio d_{0H} / d_{0Z} (Fig.4.2). Miller and Neuberger (1945) used the following Eqs.(4.1.1) and (4.1.2). Denote the bisecting point as M. Then, arc ZM = arc MH means

$$\omega = \frac{\Omega}{2}$$

$OH = R \sin \Omega$

$OZ = R - R \cos \Omega = R(1 - \cos \Omega)$

Hence,

$$\frac{OH}{OZ} = \frac{\sin \Omega}{1 - \cos \Omega} = \cot \frac{\Omega}{2} = \cot \omega \qquad (4.1.1)$$

The half-arc angle α is related to the angle ω as follows.

$MS = R \sin \omega$

$OS = \overline{O}S - R \cos \Omega = R(\cos \omega - \cos 2\omega)$

Hence,

$$\tan \alpha = \frac{OS}{MS} = \frac{\cos \omega - \cos 2\omega}{\sin \omega} \qquad (4.1.2)$$

The angle β is easily related to ω. From $\Delta H \overline{O} O$, $\gamma = \pi/2 - \Omega$, and $\beta = \pi/2 - \gamma$. Hence,

$$\beta = \Omega = 2\omega \qquad (4.1.3)$$

Table 4.1 Results of bisection of the sky

	Daytime		Twilight	Clear Night	
	cloudy	clear	clear	moon light	moon less
α^o	30 ~ 21	34 ~ 23	32	37 ~ 27	40 ~ 30
OH/OZ	2.4 ~ 3.7	1.9 ~ 3.3	2.1	1.6 ~ 2.7	1.4 ~ 2.4
β^o	45 ~ 30	56 ~ 33	51	64 ~ 41	70 ~ 46

Table 4.1 shows OH/OZ and β for values of α given in Fig.4.1. The value of α for the clear twilight sky was taken from Table 1 of

Neuberger (1951). Obviously, these values of OH/OZ and β are meaningful only for the form of sky assumed in Fig.4.2. For instance, Filehne (1912) insists that the curve shown in Fig.4.1 is a quarter of a slightly rotated ellipse, because β obtainable under the assumption of shifted circle (Fig.4.2) is too acute for his observation. If the curve is assumed to be a quarter of an ellipse, β in Fig.4.1 should be 90°. According to Filehne, this assumption was proposed by Kämtz in 1836 and criticized by Reimann in 1890 for the reason that β appeared less than 90°. Filehne came to the conclusion that the curve in Fig.4.1 is a quarter of an ellipse slightly rotated in which β is less than 90° but not too acute.

Because VS is dynamic (VS5 in Sec.1.1.1), visual distance d_0 to the sky depends upon what other percepts are visible in that direction. In the zenith direction, d_{0Z} depends only upon the condition of cloud. The decrease of α according to cloudiness from 30° to 25° in daytime in Table 4.1 may be mainly due to the decrease of d_{0Z}. When cloudy, the sky is not homogeneous and the appearance changes from aperture color mode to surface color mode. According to my own experience, the perceptual distance to the horizon, d_{0H}, is critically affected by what we see in that direction. In all the articles referred to above, I could not find any description about the terrain and other conditions in which the bisection was performed. According to these conditions, the sky may appear as a shifted circle or as an ellipsoid. Furthermore, as pointed out by Filehne, how the angle β in Fig.4.1 appears depends upon where to fixate and how to observe the horizon. When a point on the horizon is fixated, β is almost 90° and the sky within the range of vision appears vertically flat. When a more extended scene is scanned, the sky appears as a curved surface that meets the ground or the ocean with an acute angle β, but not much less than 90°.

The reason that the history of attempts to determine the shape of sky was explained above is to show that all arguments assume Euclidean structure in the sky. It is an interesting question to ask how the half-arc angle α and the angle β by which the sky and the ground meet at the horizon are related if the sky is generated in VS according to Riemannian geometry of $K < 0$.

Suppose that the same structure as shown in Fig.4.2 holds in EM, and the bisection is made as shown in Fig.4.3. I carried out an experiment to determine the shape of the night sky by asking the S to assess perceptual distances between the designated stars P_{Vi} and P_{Vj} (Sec.4.1.3). The S's had difficulty in seeing δ_{ij} as an arc along the surface of sky as the vault. Rather, they felt δ_{ij} as the chord ρ_{ij} between P_{Vi} and P_{Vj}.

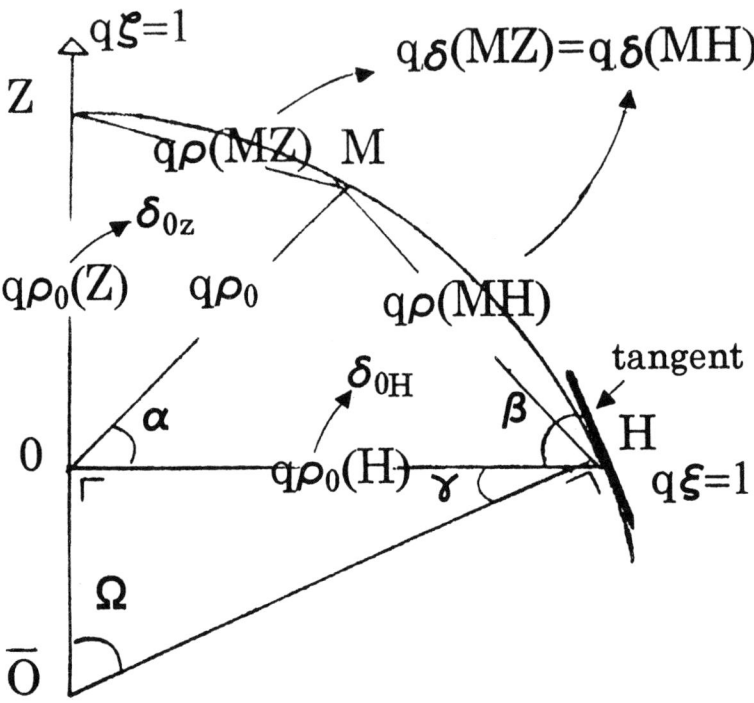

Fig.4.3 A shifted circle in EM

Let us assume that the bisection means that $\delta(ZM) = \delta(HM)$. The perceived distance δ of the chord ρ is given in Eq.(2 2.1). Then,

$$\sinh^{-1}\frac{\rho(ZM)}{G(Z)G(M)} = \sinh^{-1}\frac{\rho(HM)}{G(H)G(M)}$$

Hence,

$$\frac{\rho(ZM)}{\rho(HM)} = \frac{G(Z)}{G(H)} = \sqrt{\frac{1-(q\rho_{0Z})^2}{1-(q\rho_{0H})^2}} \qquad (4.1.4)$$

From Eq.(2.2.3)

$$R(HZ) = \frac{\delta_{0H}}{\delta_{0Z}} = \frac{\tanh^{-1} q\rho_{0H}}{\tanh^{-1} q\rho_{0Z}} \qquad (4.1.5)$$

As in Fig.4.2,

$$\frac{OH}{OZ} = \frac{q\rho_{0H}}{q\rho_{0Z}} = \cot\frac{\Omega}{2}, \text{ and } \beta = \Omega$$

Taking standard values $\alpha = 21°$ and $R(HZ) = 2.5$ in Table 4.1, let us consider a case that $q\rho_{0Z} = 0.6$ and $K = -0.88$ ($q = 0.5\sqrt{-K}$). Then, Eqs.(4.1.4) and (4.1.5) hold when $\beta = 65°$. The arc from Z to H in Fig.4.3 represents the sky in VS that is neither a part of shifted circle nor a quarter of ellipse. Since EM and VS are conformal (EM2 in Sec.2.2.2), however, we can say that the sky meets the ground or the ocean at the horizon with $\beta = 65°$. This is a more acceptable value of β than $30°$ that Eq.(4.1.2) gives for $\alpha = 21°$. Because there is no logical basis to assume in EM the form as shown in Fig.4.3, the above discussion is only for showing how the Euclidean assumption of VS prevails and the necessity to revisit the bisection of the sky from a different viewpoint other than Euclidean geometry.

4.1.2. *The Moon Illusion*

Everybody has had the experience of seeing that the moon appears much larger when it is at the horizon than when it is at the zenith. This perceptual phenomenon has nothing to do with atmospheric optics. Photographs of the moon in different positions yield no measurable difference. The moon subtends the same visual angle (about 0.5 degree) throughout from the horizon to the zenith, and that the ray from the horizon moon passes through the thicker layer of air than the ray from the zenith moon does not affect the size of the retinal image of the moon. That the moon appears larger over the horizon than at the zenith is generally called the *moon illusion* in the sense that both are generated from the retinal image of the same size. According to Plug and Ross (1989), the earliest mention of the moon illusion dates from 7th century BC or earlier, and the scientific study of the moon illusion is as old as science itself. Both originated during the period, 600 to 300 BC. The same phenomenon occurs with the sun that subtends almost the same visual angle with the moon. However, the illusion is much more noticeable with the moon, and many experiments with the natural moon as well as with a simulated moon have been conducted and many "theories" or "hypotheses" to "explain" the moon illusion have been proposed (*e.g.*, Hershenson, 1989).

It is natural to think that the moon illusion is related to the flattened shape of the sky. When one fixates for a while on a bright square on the dark background and then turns the eyes to a vertical white plane, one sees the dark square on the plane. This is called the (negative) after image. As the distance from the eye to the plane is changed, the size of the after image proportionally changes in the neighborhood of the body (Fig.4.4A). Denote by α, D, and s the visual angle of the fixated square, the distance and the length of after image. Then, s is proportional to αD,

$$s \propto \alpha D$$

This is called the Emmert's law. Suppose that the same relation holds between the retinal image of the moon and the perceptual distance *d* to the surface of the sky at which the perceived moon of size *s* is localized, then the moon illusion can be considered as a special manifestation of Emmert's law ($d_H > d_V$ in Fig.4.4B). This is called the apparent-distance hypothesis or the size-distance-invariance hypothesis, often abbreviated as SDIH ($\alpha \propto s/d$). This explanation of the moon illusion is often credited to Ptolemy (ca.142 AD). According to Ross and Ross (1976), however, Alhazen (Latin name of the Arabic scientist Ibn al-Haytham, c.965-c.1040) was the first to explain the moon illusion by the SDIH. The figure illustrating this hypothesis given in 1738 by R. Smith, an astronomer in Cambridge, is widely cited. Fig. 4.4B is a modified form of that figure. As to methodological problems, *i.e.*, how to compare the moon at different positions and how to quantify the perceived size as s_H and s_V, consult Kaufman and Rock (1989).

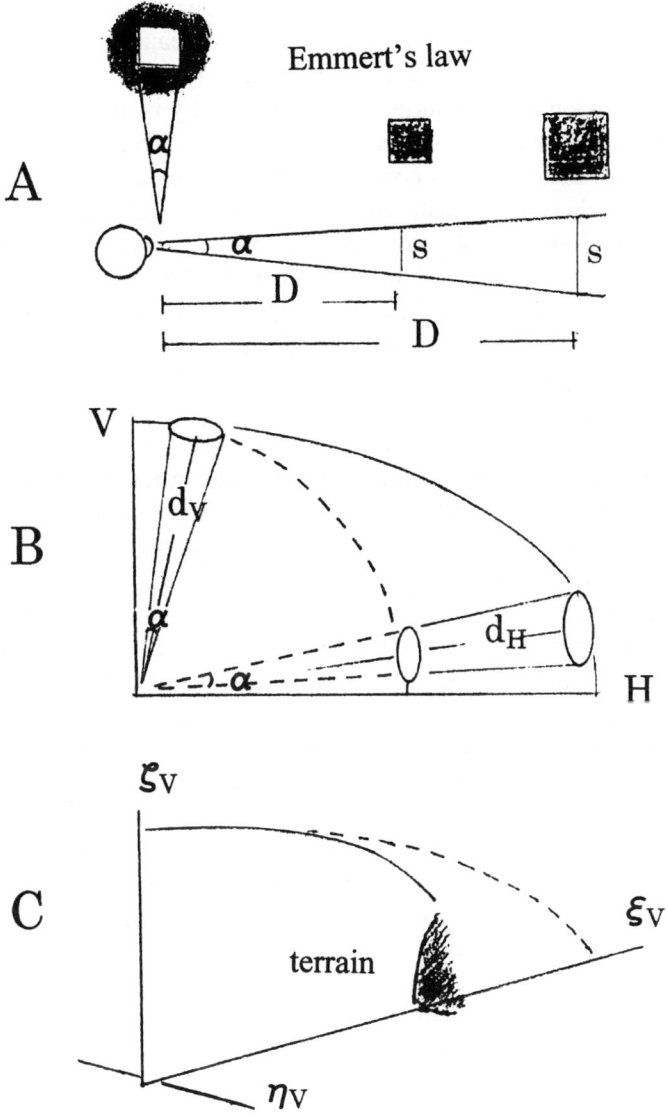

Fig.4.4 Emmert's law and moon illusion

The difficulty of this explanation is that the moon at the horizon does not necessarily appear far away as depicted in Fig.4.4B. Most people judge the horizon moon to be closer. Some investigators ascribed this judgment to the fact that the moon appears larger. Plug and Ross (1989) called this paradoxical situation "further-larger-nearer theory" in the sense that the horizon moon appears larger because it is far, but the moon appears nearer because it is larger. In my own experience, when the horizon moon appears large and close, the sky is more curved before it reaches the horizon. As stated in VS5 in Sec.1, VS is dynamic and VS is structured differently according to what are visible in the given direction. If there is no distinctive pattern in the light coming from the entire physical sky, the sky appears as a flattened surface that meets the horizon with an angle less than $90°$. When the horizon is obstructed by terrain, the sky right above the terrain appears either at the same distance with the terrain or at an indefinite distance from the terrain as its background (Fig.4.4C). This part of sky is not the extension of the sky being flattened from the zenith (dotted curve in C). Baird and Wagner (1982) reported a case in which students judged that the distance to the sky, right above a building situated 300m away at the end of the road in the Dartmouth campus, is less than the half of the distance to the zenith. How VS extends in a direction depends upon the stimulus configuration in the physical space X. The perceived sky is not a rigid surface to which many percepts, *e.g.*, the moon and the terrain, are pasted without affecting its form. Either the terrain visible in the direction of horizon or the presence of the moon itself can alter the shape of that part of the sky.

The purpose of this section is not to review many experimental results and theories concerning the moon illusion but to point out that the moon illusion should not be treated as an isolated problem of its own right. The moon and the sun subtend almost the same visual angle of about $0.5°$. Both look very small when taken in a photo. In direct observation, both appear as if they are objects subtending a much larger visual angle. The artificial moon at the Hayden planetarium projects a $1°$ image on the dome and still "many people are disappointed being of the opinion that it appears too small." (Gilinsky, 1980). Fig. 4.5 shows the retinal image at a single glance of the moon at different positions in the sky. The image of moon is a tiny dot. Although the retinal image is inverted, for the sake of convenience, they are shown upside down. When the moon is at the horizon, the lower half is occupied by the image of the ground (A).

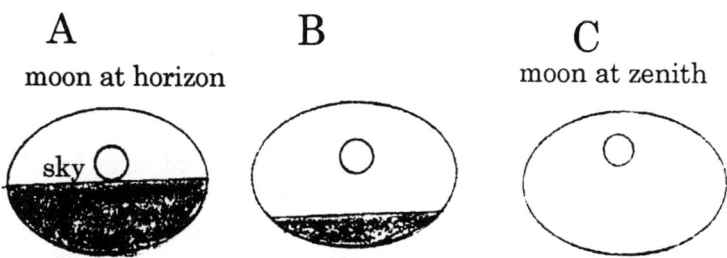

Fig.4.5 Images of the moon on the retina (upside down)

From this pattern of proximal stimulation, VS is generated in which the sky and ground stretch in the direction of ξ_V to meet at the middle. The distance δ_{0H} to the horizon and the size s_H of the horizon moon are determined by the pattern in the ground image, because the sky image is homogeneous. When the moon is in the middle sky and the line of sight is moved upward, the ground image becomes smaller (B) and when the zenith moon is being fixated, the entire retina is occupied by the homogeneous sky image with the tiny image of the moon at the center (C). From this pattern of proximal stimulation in the elevated eyes, the zenith sky at δ_{0V} and the zenith moon of size s_V are generated in the direction of ζ_V. As this pattern C is almost the same with Ganzfeld condition (VS5 in Sec.1), it will generate a different VS if glances of other parts of sky are not accompanied. The shape of the sky and the moon illusion are the product of multiple glances. What is meant by the moon illusion is that $s_V < s_H$. The question of why the moon illusion takes place should be embraced by the more general question of why VS is structured as it is. Gilinsky (1980, 1989) emphasized the same idea and proposed adaptation theory of visual space.

4.1.3. *Multidimensional Construction of the Night Sky*

Stars are perceived as being embedded in the night sky as a vault. Using a set of stars as $\{Q_i\}$, we can construct, in a 3-dimensional Riemannian space R^3, a configuration of points $\{P_i\}$ that represents the night sky we are perceiving. The methodology was described in Sec.3.2.1. Three

114 *Global Structure of Visual Space*

experiments were performed with simulated stars in a dark laboratory room, and one outdoor experiment was performed with real stars (Indow, 1968). Herein only the results of the outdoor experiment are described.

Stars were observed in autumn 1962 at Oiso Beach by two S's, T.I. and T.K. This is a resort beach not too far from Tokyo and has a long coastline that is fairly free of stray lights from towns. Eleven stars selected as Q_i, and $\{P_i\}$, locations of the perceived stars in EM, are shown in Fig.4.6. Stars of about the same brightness, each being clearly distinguished from surrounding stars, were selected from the entire sky. The S was included as Q_0 and hence $\{Q_i\}$, $i = 0, 1, 2, ..., 11$. The S's sat on the shore facing the sea of Sagami Bay in the direction of W20°S, and made ratio assessments as shown in Fig.3.6 on perceived distance δ_{ij}. One session was limited less than one hour to keep the shift of the star configuration within a tolerable range. Four sessions were needed to obtain the complete distance matrix data $\mathbf{D} = (d_{ij})$ based on two repeated assessments (n = 2). Each session was at night with clear sky without the moon.

As mentioned in Sec. 4.1.1, it was not easy for the S to sense δ_{ij}, $i \neq j$, along the curved surface of the sky. Hence, what is meant by δ_{ij} is, not the length of arc, but the length of chord between perceived stars. This is the distance we need as the input data to MDS. When a star was intensely fixated, it tended to recede into the sky farther and farther. The S was recommended to move the eyes quickly between three stars (Q_i, Q_j, Q_α) to respond with $r_{i,ja}$ in Eq. (3.2.1). Some stars, *e.g.*, 5 and 6, were almost along the ζ-axis and some stars, *e.g.*, 1 and 10, were in the η-axis direction. The S had to scan, by moving his head but without turning the body, the entire sky visible by this posture. The impression of vault form of sky was vivid. To construct $\{P_i\}$, the data matrix $\mathbf{D}$ was first processed by the metric MDS (Indow, 1968, 1974) and then by DMRPD (Indow 1990, 1991).

In DMRPD, the physical location of stars in Fig.4.6 was used as the initial configuration $\{P_i\}^{(0)}$ with an arbitrary scale. Theoretical distances $\hat{d}_{ij}$ were determined from interpoint distances ρ_{ij} in $\{P_i\}$ under the assumed sign of K (Eqs.2.2.1 – 2.2.3). The coincidence between $\hat{d}_{ij}$ and data d_{ij} was evaluated by Stress (Eq.3.2.11). The finally adopted configuration $\{P_i\}$ was re-scaled so that the mean of $\rho_{0i} = 1.0$.

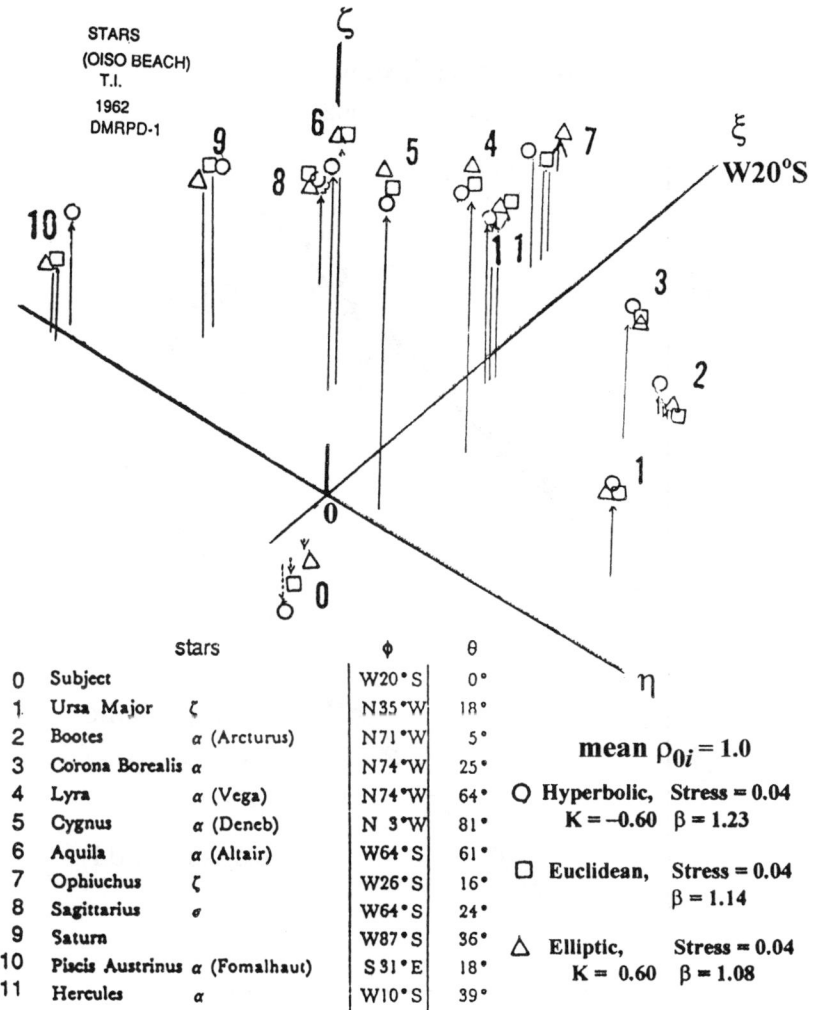

Fig.4.6 Configuration of stars constructed by DMRPD

The values of K are in terms of this unit. Three $\{P_i\}$'s, $K < 0$, $K = 0$, and $K > 0$, were replotted in Fig.4.6 in a form suitable to this book. The scatter diagrams, d_{ij} vs. $\hat{d}_{ij}$ of the same S for K = 0 and K < 0 are shown in Fig.4.7.

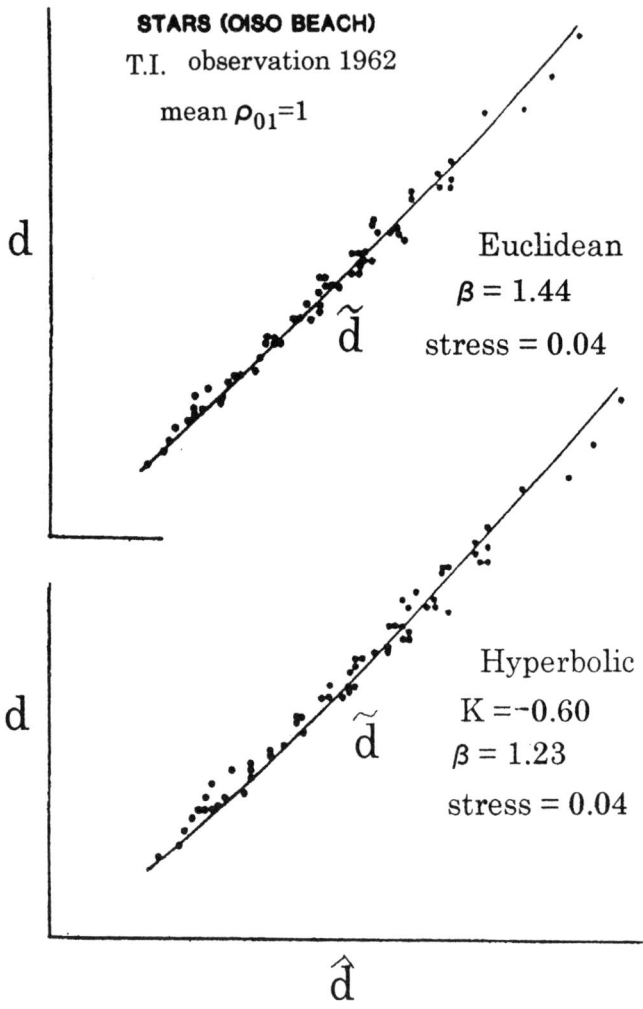

Fig.4.7 Two examples of the relation between data d and $\hat{d}$ in Fig. 4.6

In contrast to $\{P_i\}_P$, $\{P_i\}_D$, $\{P_i\}_{H\alpha}$ in the horizontal DP, it is difficult to decide which geometry is most appropriate to represent the perceived configuration of real stars. The three $\{P_i\}$'s are almost indistinguishable, the values of Stress and the forms of $\tilde{d}(\hat{d})$ are almost the same. Stars are a 2-D configuration embedded in the sky surface and there is nothing in between the self and stars. The sky can be regarded as a surface we would have when a frontoparallel plane HP is moved up to the boundary of VS. The S scanned the sky by moving the head to observe stars, which means that the part of sky being observed is always perpendicular to the line of sight. Then, it is not surprising that the configuration of 11 stars is represented equally well by $\{P_i\}$'s structured in three different geometries. We encountered the same situation in representing Hh- and Dh-alleys on an HP (Fig.3.10). In that case, however, it was concluded that the HP is of $K = 0$ because of the coincidence between Hh- and Dh-alleys (Fig.3.5). In the present case, some information other than **D** is needed to pinpoint K of the sky.

Putting aside the problem of geometry, it is clear in Fig.4.6 that distance d_0 from the self to the sky coplanar with the stars is different according to direction. Let us denote by d_H, d_Z and d_S distances to the sky in the horizon (ξ), in the zenith (ζ), and in the left or right η-direction. Then, d_H is the shortest, d_Z is slightly larger than d_H, and d_S is considerably larger than the other two (see Fig.5 in Indow 1968). This is in agreement with the impression that the S had who was sitting on the seashore. As emphasized at the end of Sec.4.1.2, how VS extends in a direction is determined by what is visible in that direction. On this seashore, the S sees nothing but the dark sea and sky in the ξ-direction. The S sees stars in the ζ-direction, and lights of houses and terrain at a distance in the η-direction.

4.1.4. *Horizon*

For one looking straight ahead, the horizon, if it is visible, always appears at the height of the line of sight. One half of the retina is covered by the image of sky and the other half by the image of ground or sea (Fig.4.5A). From this proximal stimulation, such a VS is generated that the eye level of the self and the horizon are located at the same height C on the ζ_V-axis (Fig.4.8). The situation is the same no matter

118 *Global Structure of Visual Space*

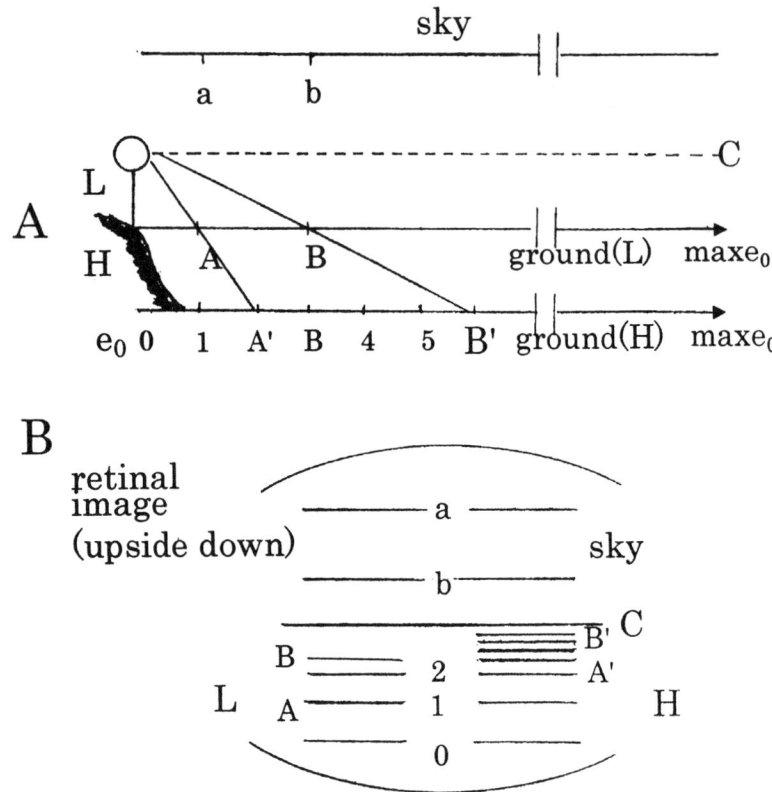

Fig.4.8 Physical condition of horizon. Two different viewing positions

whether one stands on the flat ground (Lower position L) or on a high hill (Higher position H). The situation will be easy to understand if we imagine such a case to stand on the broad road that stretches straight ahead. The left and right edges of the road appear to meet at a point. This is called the vanishing point in the perspective. The figure A is equivalent to its 90° rotation; the sky is one edge of the road and the ground is the other edge. The figure B shows which part of the ground is imaged on which position of the retina with regard to the line C (upside down). The left is when one is at the lower position L. The right is when one is at the higher position H, where the image of the ground from the same distance e_0 from the body is more shifted toward the central line C and images are more congested. The retinal image of the sky will not change much according to the standing position.

Let us consider which part of the ground corresponds to the horizon in the schematic illustration of Fig.4.8. Denote by e_{0i} the physical distance to a position i on the ground and by δ_{0i} the perceptual length of the corresponding radial distance (Fig. 1.3). As pointed out in VS3 in Sec.1.1.1, $\delta_0 = f(e_0)$ has the asymptote max $\delta_0 = f(\max e_0)$ in a given direction. For $e_0 > \max e_0$, δ_0 remains at max δ_0 and hence the image of max e_0 must be at the line C in Fig.4.8B. If max e_0 is determined by the density gradient of the image of the ground, max e_0 is shorter for one at the standing position H. If we look down the ocean from a high hill, the horizon appears closer. It is especially impressive when there is an island beyond the horizon.

We can find many excellent quantitative descriptions of perceived forms of real scenery in books of a geographer Cornish (1935, 1937). He invented and practiced to "delineate as much of the arc of the horizon as constituted the field of his conscious vision filling the page of his sketchbook". His head was fixed and only his eyes were moved. The viewing position was known and the features were identifiable on the map. Then he compared the angles subtended in the sketch with the actual angles in the map. In this way, he discussed the appearance of the horizon at Riviera and other places. As I do not have an aptitude of sketching, I will present below only schematic illustrations.

Suppose that one is on the high position of a slope road going down to a cliff under which there is a beach (Fig.4.9A). The view of the sea is delineated by two hills on both sides of the road and the texture of the sea surface is not perceptible. Then, the sea and the sky appear as parts of the plane filling the gap between the hills. In other words, the sea looks like a blue wall vertically standing and the horizon is straight from left to right. Someone just cannot believe that the wall is the sea. When one comes down to the top of cliff, one has a wider view of the sea. The sea extends in the ξ_V–direction but much more along the η_V–axis (Fig.4.9B). If the terrain is not visible, the horizon is straight, keeping the same height from left to right with regard to ζ_V–axis. When one comes down to the seashore, the sea appears as the tilted plane, from $\zeta_V(0) = 0$ to $\zeta_V(0) = h$ (the height of person) along the ξ_V–axis ($\varphi_V = 0$), and the horizon is a curve connecting the left and right terrains on the

η_V–axis (Fig.4.9C). The length of $\delta_0(\phi_V)$ is determined by various factors. The radial distance in the straight-ahead direction, $\delta_0(\phi_V=0)$, may be determined by the gradient of texture density of the sea surface, whereas δ_0 at $\phi_V = \pm 90°$ is affected by the perceptual distance to the terrain, which is the end of a continuous succession of various percepts on the ground. The presence of continuous succession of objects on e_0 makes larger max e_0 and hence max δ_0.

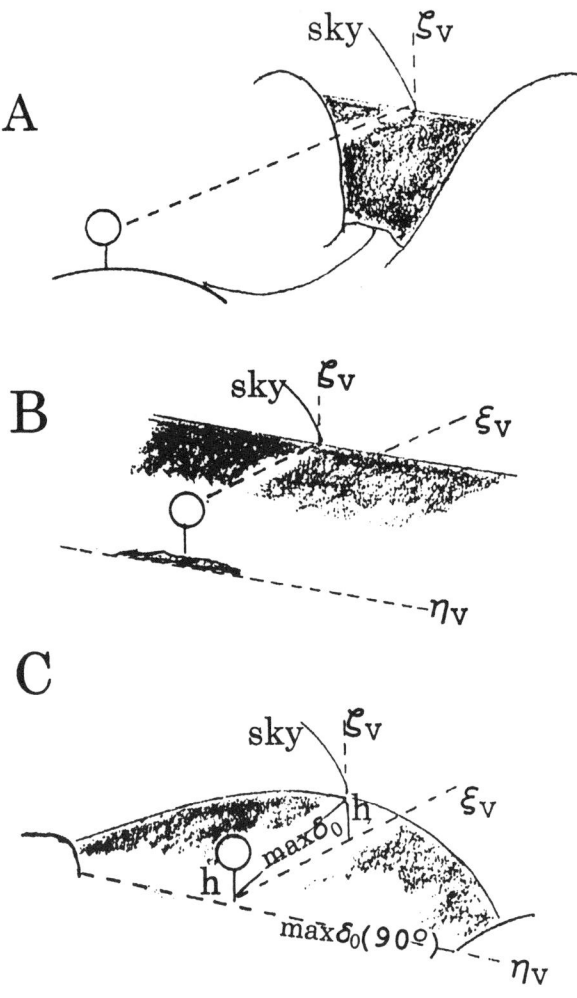

Fig.4.9 Appearance of sea

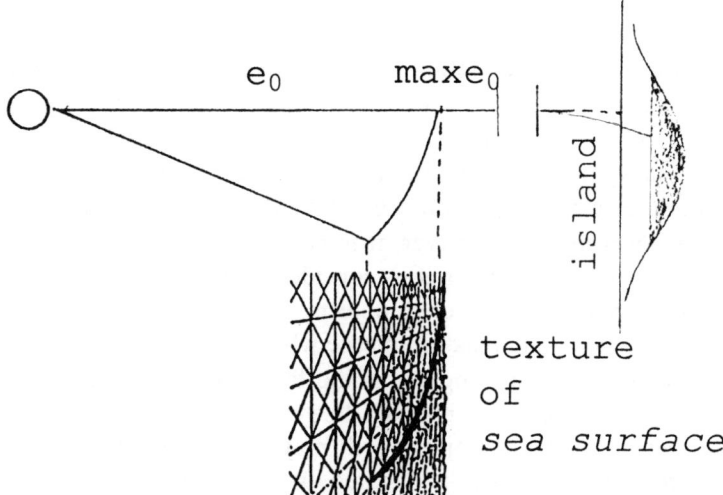

Fig.4.10 Horizon of sea

In the direction of ξ_V, what changes along e_0 is only the texture density of the sea. The intermediate $\delta_0(\varphi_V)$ may be functions of the gradient of texture density of the sea surface in that direction as well as $\delta_0(90°)$ and $\delta_0(-90°)$. The locus of constant gradient of texture density of the sea surface is curved according to direction $\pm\varphi_V$ (Fig.4.10). Often, the height of horizon $\zeta_V(\varphi_V)$ also changes according to φ_V from h at $\zeta_V(0)$ to $\zeta_V(\pm 90°)$ = almost 0 (Fig.4.9C). Someone ascribes this curvilinear appearance of horizon in the ζ_V–direction to the fact that the earth is round. It is not the case, however. If the visual span is delineated and the visibility of terrain in the directions $\varphi_V = \pm 90°$ is excluded, the horizon appears straight as in Fig.4.9B. As shown in Fig.4.10, if there is an island at a far distance on the line of sight, we can see only its upper part, which is certainly due to the curvature of the earth. An example of the appearance of horizon when viewed from a ship was shown in Indow (1999). At any rate, the appearance of natural outdoor scenery is intrinsically related to $\delta_0(\varphi_V)$. There are several studies that try to make explicit how δ_0 changes as a function of e_0.

4.2. Scaling of Radial Distance δ_0

Most studies dealt with only $\delta_0(0)$ in the straight-ahead direction ($\varphi = 0$) as a function of $x = e_0(0)$. In this section, the perceived and physical radial distances in this direction are respectively denoted as δ and x. In order to relate δ to the physical radial distance x, we have to convert the latent variable δ to a scaled value d. Several studies will be reviewed from the viewpoint of methodological problems in this scaling. If we take a point at a distance x on the sea in Fig.4.10, it is perceived at a distance of $\delta(x)$ in VS. However, it is not easy to discuss the case of a continuous surface like sea. Let us consider the case in which there are a continuous succession of discrete objects, $i = 1, 2, \ldots$, on the x-axis in X between the body and max e. In case when an object i is ego-centrally localize, the convergence angle γ_i is the factor to determine δ_i and $\delta(\gamma_i)$ was discussed in previous chapters. In general, however, the condition surrounding the series of object i will play a more fundamental role in determining how far each is located in VS. Hence, what is to be discussed in this chapter is the functional form $d\{\delta(x_i)\}$.

4.2.1. Scaling based on Difference Judgment

Gilinsky (1951) published the results scaled by a method that she called a modified form of the *method of equal-appearing intervals*. The S stood at one end of an unfamiliar indoor archery range, about 80 ft long. The experimenter moved a pointer stick at a slow and nearly constant rate along the ground away from the S to mark off successive increments that appears to be the length of one foot. Each successive increment was regarded as an attempt to match, by non-simultaneous viewing, a memorized "subjective foot rule". An increment point was temporarily marked by a marker. As soon as the next setting was attained, the marker was moved to that increment point. Namely, when the increment point i was judged, the S could see nothing but the pointer stick and the marker of x_{i-1}. She defined the perceptual distance $d(x_i)$ to be i feet in terms of the subjective foot rule. Regarding this foot rule and the physical foot rule to be commensurable, she expressed both d and x with the unit of ft, d(ft) and x(ft). She plotted the data by taking discrete values of $d = 1, 2, ...,$ ft on the ordinate and the corresponding x(ft) on the abscissa and fitted the curve

$$d = \frac{D \times x}{D + x}, \quad x > 0, \quad D > 0 \tag{4.2.1}$$

The only one parameter D determines the deviation of the curve $d(x)$ from the curve $d = x$. The series of points in Figs.4 (S: AG) and 5 (S: JB) in her article are closely on the curve (4.2.1) in the range $0 < x < 80$ ft. As $d(d)/dx < 1$, the curve is concave downward.

In contrast to x(ft) in terms of physical unit ft, the variable d is the result of successive matchings with the subjective one foot rule. Hence, let us keep the units of the two variables separate and denote the latter as d(sft) and consider

$$d = D\left(\frac{x}{M + x}\right) \tag{4.2.2}$$

where D is given with the subjective unit by which d is defined and M is a parameter in terms of the physical unit by which x is defined. Eq.(4.2.1) with d(sft) is a case of Eq.(4.2.2) in which M for x(ft) has the same numerical value with D. The equal-appearing interval scale d in Eq.(4.2.2) can be expressed in terms of any unit such as sm when each interval is supposed to match the subjective length of the physical 1 m.

Harway (1963) repeated the same experiment in a level grassy field with S's of five age groups (mean ages of 5.5, 7, 10, 12 years old, and adult). Each group consisted of 12 S's. The open terrain in front of the S extended for at least 65 ft without any clear markers such as bushes. The standard one-foot rule was placed on the ground at the front of and immediately to the foot of the S. The S directed the experimenter to mark off successive 1-sft intervals comparing with this standard. When the increment point i was judged, the S could see the pointer stick, that was to be moved by the experimenter, the yellow marker at x_{i-1}, and the standard one-foot rule. He called Δ_i = mean of $(x_i - x_{i-1})$ over S's "constant error" in the sense that Δ_i being less than 1ft means the difference between subjective and objective one-foot lengths. What he showed was the plotting of Δ_i against i. Since the mean height of eyes from the ground is different according to age, he conducted the experiment under two conditions, "normal height" and "adjusted height". In the latter, a child stood on a platform of an appropriate height and an adult kneeled so that the height was fixed at about 5.5 ft for all S's. Under the normal height condition, 12 years old children exhibit the same result as adults.

From the smoothed curves in the Fig.1 of the original article, I defined curves $d(x)$ that correspond to the plotting in the Gilinsky's article. The ordinate is d on which i varies, 1, 2, ..., and the abscissa is x_i = the sum of $\Delta_{j.}, j \leq i$. Fig.4.11 shows $d(x)$ thus defined for adults and 7 years old children under the normal height condition. The curve of Gilinsky's Fig.4 (S: AG), the broken curve of Eq.(4.2.2) with D = 94 sft and M = 94 ft, is included.

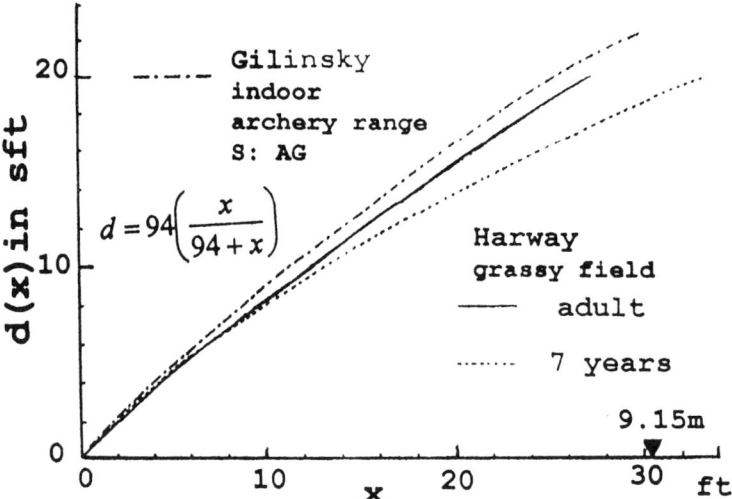

Fig.4.11 Scaled distance $d(x)$ by equal-appearing intervals in two studies

One problem with the scaling procedure in these two studies is that random error at the j-th increment is carried over to all the subsequent results $d(x_i)$, $i \geq j$. It is not essential how 1 sft is related to 1 ft. It is critical, however, that the following condition is satisfied

$$\varepsilon = [\delta(x_i) - \delta(x_{i-1})] = \text{constant} \tag{4.2.3}$$

where ε can be in any unit. It will be necessary to try a verification observation of this condition with the same S's under the same condition. If markers are placed at all the increment points (if they are too many, then at $i = 2, 4, 6,..., etc.$), the series should appear to be of

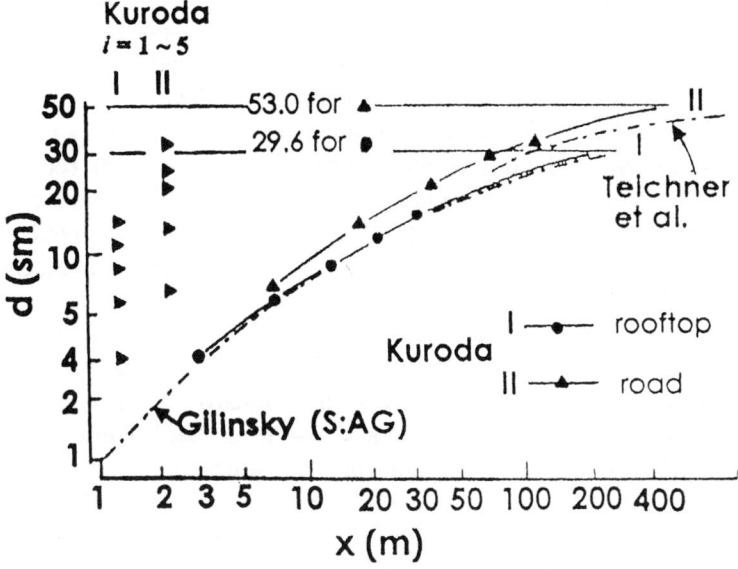

Fig. 4.12 Distance functions by equal-appearing intervals on log-log coordinate

equal-intervals. This caution was taken in a study by Kuroda (1971) in which he obtained a series of successive increment points $i = 1, 2, ... , 5$, (I) on the rooftop of a building (47.5 m long and 13 m wide) and (II) on a road (127 m long and 6.5 m wide) with the same five S's. The standard physical length was 3 m in the former and 7 m in the latter. The S was allowed to re-adjust the distances between markers after all five markers were set. Each S repeated settings four times. He fitted Eq. (4.2.1) to x_i (in m), the mean results over S's. I converted this scale of d to the unit sm (the subjective size of 1 m) and plotted the results in the log-log coordinates (Fig.4.12). Points (rooftop) and triangles (road) respectively jump on the ordinate with intervals of 3 sm and 7 sm (allows in the Figure). Converting x(ft) to x(m) and d(sft) to d(sm), I included the Gilinsky curve in Fig.4.11. This curve closely represents the adult curve of Harway also. The horizontal lines represent D's of the respective curves in the sm unit. The parameter in Gilinsky, D = 94 ft, is converted as 28.7 sm. The value of D given by Kuroda for the rooftop curve (I), 9.88 in terms of 3 sm step, and that for the road (II), 7.57 in terms of 7 sm, are respectively converted to 28.7 sm and 53.0 sm. According to the series of black triangles, D = 53 sm roughly corresponds to $x = 700$ m, which implies that, if this road is extended enough, the scene of the road at about 700 m ahead appears as D and 53 sm may be the boundary of VS in this direction under this condition. The curve labeled "Teichner et al." is explained below.

Teichner et al. (1955) performed an experiment on a very extensive field, an airstrip in the desert near Yuma, Arizona. A black rectangular board, 77 in. high and 64 in. wide, was placed at x_i as the standard target. On its left side, a variable target of the same size was presented with a separation of visual angle of 3 min. The variable target was attached to the back of a jeep and moved at almost invisible speed from either "far" or "near" side toward the standard until two targets appeared to be at the same distance (the method of limits). Four soldiers served as S's and each repeated judgments 6 times with the standard at each x_i. Hence, 24 settings were obtained from "near" and also from "far". These values are distributed around x_i and the standard deviation of the combined 48 settings is defined as the differential threshold $\Delta(x_i)$. Position x_i of the standard target was varied 9 ways from 200 to 3000 ft. They obtained $\Delta(x_i)$ under two conditions, binocular and monocular observations. Both observations gave essentially the same results, which means that the binocular cues γ in Fig.1.3 played no role for these distances. Fig.4.13 shows $\Delta(x_i)$ under binocular observation re-plotted in a slightly different way from Fig. 2 of the original article. Excluding two white circle points at 1500 and 2200 ft, they fitted a power function,

$$\Delta(x) = ax^b, \quad a = 0.0006 \text{ for } x(\text{ft}), \quad b = 1.157 \qquad (4.2.4)$$

They wrote "all other distance locations were level but these two distances lay on a slight rise or slope of approximately one to two degrees". They interpreted that these two $\Delta(x)$'s were small as the artifact due to these circumstances, and later confirmed in an indoor experiment that, if the targets were located on the rising ground, the S's sensitivity to differences in target-distance was increased, which means that $\Delta(x)$ becomes small (Dusek, et al., 1955).

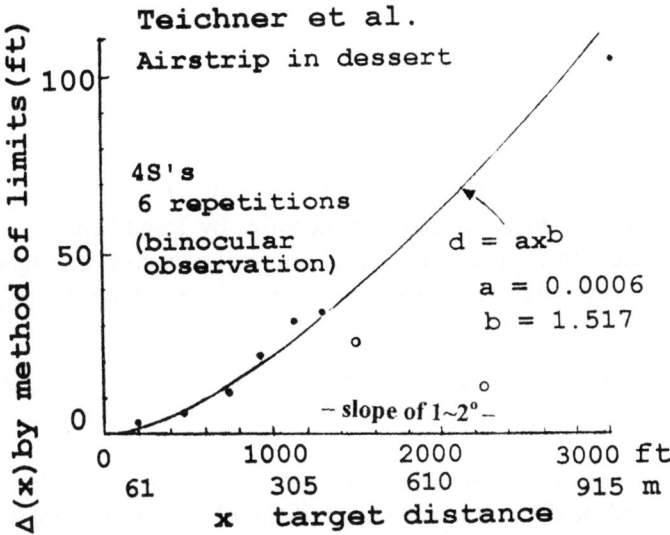

Fig.4.13 Differential threshold $\Delta(x)$, jnd, at x (Teichner *et al.*)

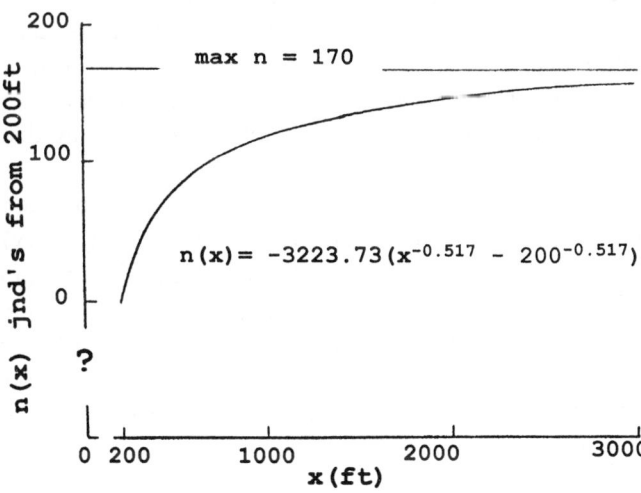

Fig.4.14 Number of jnd's between x and 200 ft, estimated from Fig. 4.13

128 Global Structure of Visual Space

In general, if a series of points $1/\Delta(x_i)$ plotted against x_i are fitted by a smooth curve between A and B, $i = 1, 2, ...$, we can define the number of $\Delta(x)$ intervening between A and B. Let us denote that number by n

$$n = \int_A^B \frac{1}{\Delta(x)} dx \qquad (4.2.5)$$

When x increases from $x_1 = A$ to $x_2 = A+\Delta(x_1)$, $n = 1$, and when x increases from x_2 to $x_3=x_2+\Delta(x_2)$, n increases by 1 and hence, $n = 2$, etc. Namely, the same procedure for defining the curves in Fig.4.11 by adding Δ_i one by one can be carried out analytically. From Eq.(4.2.4),

$$n(x) = \frac{1}{a}\int_A^x x^{-b} dx = \frac{-1}{a(b-1)}\left[x^{-(b-1)}\right]_A^x \qquad (2.4.6)$$

Fig.4.14 shows $n(x)$ obtained from (2.4.6) where A = 200 ft, the smallest value of x of the data. It is not known if Eq. (2.4.4) strictly holds for $0 < x < 200$ ft (61 m). Though the curve (2.4.6) does not have the asymptote, if the curve in Fig.4.14 is expressed by a function having max n, its value may not be far from 170. Another name of differential threshold $\Delta(x)$, is "just-noticeable difference, jnd." of stimulus. Namely, if a stimulus of an intensity x_i is increased, then the S is expected to detect the change when the intensity reaches $x +\Delta(x)$. Hence, starting from $x = 200$ ft, about 170 jnd steps exist in this direction under this condition. From this result, no information is obtained as to how many steps we can discriminate between the body and 200 ft.

Suppose that all increments in the perceived distance δ that corresponds to $\Delta(x)$ are of the same size, $\varepsilon = \delta(x + \Delta(x)) - \delta(x)$. Then, $n(x)$ can be regarded as the scale $d(x)$ that represents $\delta(x)$ with the unit of ε. This has been a problem in dispute in psychophysics for a long time. It might be of interest to try to connect the curve in Fig.4.14 to the curves in Fig.4.12 under the assumption of ε being a constant. It can be done if we add some assumptions on the value of C = $n(200\text{ft})$ and on the size of ε. Using $x(m)$ and assuming C = 600, we can express the curve in Fig.4.14 as

(d in terms of ε) = $-5956.3\,(x^{-0.517} - 0.1194) + 600$

If we assume that 600 with the unit of ε is equivalent to $d(\text{sm}) = 25$, we have the curve

$d(\text{sm}) = 0.042\,(d$ in terms of $\varepsilon)$

This is the curve of broken-line from $x = 61$ m (200 ft) to 900 m labeled as Teichner et al. in Fig.4.12. This is not the result of optimization. If this conversion is taken at its face value, we can say that 24 jnd steps are in an interval of 1 sm under this condition. It is known in general that, if scales $d(x)$'s of equal-appearing intervals are defined for the same $\delta(x)$ with different units, these $d(x)$'s coincide with each other well if the differences between units are adjusted. The curves in Fig.2.12 show more than that. Scales $d(x)$ for radial distance $\delta(x)$ defined under different contexts exhibit more or less the same form except the level of D. It is understandable that the level of asymptote changes according to the context, e.g., D for x on the road is larger than that for x on the rooftop.

4.2.2. Scaling based on Ratio Judgment

To scale δ, many investigators used the method called *magnitude estimation*. This procedure was promoted by Stevens (1956, 1975) and has been applied to scale sensory intensity in many modalities. Let us denote by s the physical intensity of stimulus and by $v(s)$ a scale obtained by magnitude estimation. It was found by Stevens that, if $v(s)$ is plotted against s in log-log coordinates, the main part of the trend is linear in many modalities. The procedure to have a scale $d(x)$ of radial distance $\delta(x)$ by magnitude estimation is as follows. A target is presented at x_i, $x_1 < x_2 < \ldots < x_N$, one by one usually in a random order, and the S is asked to assign a numerical value z_i to x_i that is supposed to be proportional to δ_i. When the evaluation is repeated n times with x_i, usually we have a skewed distribution of n values of z_i. Hence, either the median or the geometric mean is taken as the representative value $d(x_i)$. It is expected that

$$\frac{d(x_i)}{d(x_j)} = \frac{\delta(x_i)}{\delta(x_j)} \qquad (4.2.7)$$

Plotting $d(x_i)$ against x_i, we can define a scale $d(x)$. It is irrelevant by what unit $d(x)$ is defined. In some studies, when assigning z_i to x_i a modulus is specified. That is, the S is told that a value (*e.g.*, 10) should be given to the first stimulus A, *i.e.*, $z(A) = 10$. Then, the unit of $d(x)$ is defined by this modulus. In some studies, no modulus is given and it is up to the S what value $z(A)$ is assigned to the first stimulus A. The S is only requested to assign the value to the next stimulus B in proportion to $z(A)$. Namely, $z(B) / z(A) = \delta(B) / \delta(A)$. Some S's are not sure about what is meant by assigning a value $z(s)$ in proportion to the sensory experience caused by a stimulus of intensity s, and it is recommendable to have some practice trials using some obvious examples. Often line lengths drawn on a sheet of paper are used for this purpose.

Once $d(x; u_0)$ is defined with a unit u_0, it can be converted to a scale $d(x)$ with another unit u, $d(x) = (u/u_0) d(x; u_0)$. Though x as the physical radial distance in VS is not necessarily the intensity of stimulus in the same sense that the intensity s of light is to brightness or acoustic intensity s to loudness, it has been found that, as is the case in $v(s)$, the main part of $d(x)$ is linear with x when plotted in log-log-coordinates. In other words, a power function holds in the main part,

$$d(x) = \alpha x^\beta, \quad \alpha > 0, \; \beta > 0, \qquad (4.2.8)$$

The slope of line in log-log coordinates is determined by the exponent β and β is independent upon the units of x and d. The other parameter α defines the value of d when $x = 1$. Often, the trend is slightly curved downward at its lower end. This fact will be discussed in Sec.4.2.3. If the unit of d is changed, the line in log-log coordinates simply moves along the log d ordinate with the same slope β.

Galanter and Galanter (1973) obtained $d(x)$ by magnitude estimation over very wide ranges of x under five conditions. Medians $z(x_i)$ of S's, who were given training in making magnitude estimations of line length, were plotted in log-log-coordinates. Results in Figs. 1, 2, 4, and 5 in the original article are summarized in Fig.4.15 as I, III, II, and IV. The scale of ordinate d is given with an arbitrary unit in each experiment. Hence, only for showing the slopes, all curves are plotted in Fig.4.15 in appropriated positions along the ordinate.

I. Twelve S's were located on a missile launch pad on a beachfront overlooking the ocean. An aircraft traversed a line perpendicular to their line of regard at various distances x_i at low altitudes (< 200 ft above the water), sometimes from the left and sometimes from the right. The slope β of the liner part is 1.25.

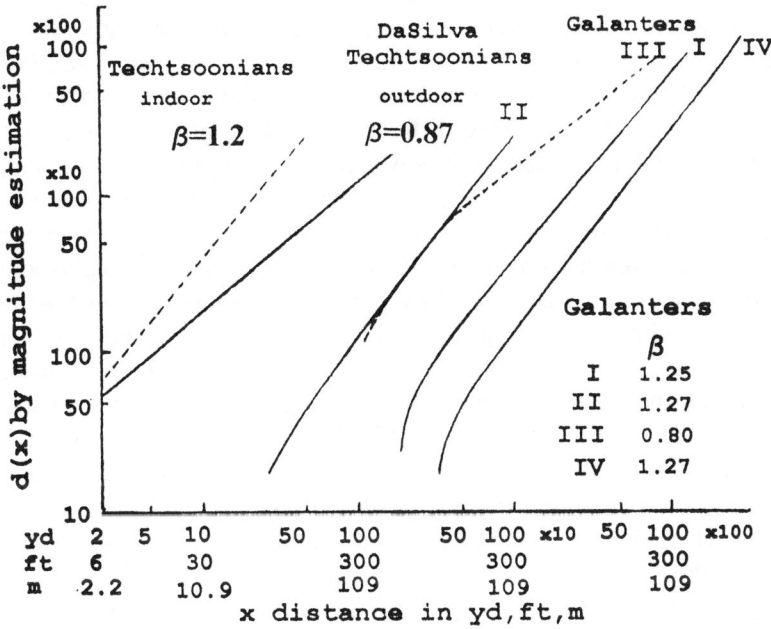

Fig.4.15 Distance scales by magnitude estimation on log-log coordinates

II. Nine S's were on the shore of Great Bay, New Jersey. A 26-ft cruiser passed, with a constant speed, perpendicular to their line of regard at varying distances x_i. The slope β is 1.27, about the same as in II. Namely, whether a target is embedded in the sky or in the ocean surface does not affect the value of exponent β.

III. A light aircraft flew directly overhead at x_i according to a prearranged irregular schedule of altitude. Eight S's made magnitude estimations with a fixed modulus ($z = 1000$ to the first flight of altitude 550 ft). The slope β for this vertical direction is 0.80, considerably smaller than for the horizontal direction.

IV. Six S's worked in groups of three. Three boarded the experimental aircraft that flew at various altitudes over a prearranged course. When targets on the ground were over flown at distances x_i, they made magnitude estimations $z(x_i)$ without a fixed modulus. The slope β for this air to ground distance is 1.27.

Da Silva and Das Santos (1982) performed magnitude estimation experiments in a large open field, 300 m long and 30 m wide. There were trees and bushes at the sides and at the far end. The target (yellow triangle) was placed at x_i that was varied 10 ways between 2 to 296 m. The size of triangle was different according to the distance. Magnitude estimation was made in three different ways with 80 S's in total. The mean slope β for 80 individual power functions is 0.87 with a standard deviation of 0.14. Similar results were obtained in a subsequent experiment (Da Silva and Da Silva, 1983). In an outdoor experiment, when x_i was varied from 5 to 480 ft, Teghtsoonian and Teghtsoonian (1970) reported that the mean slope β for 48 individual power functions was 0.87 with a standard deviation of 0.13. They did the same experiments in two natural indoor situations, one in a large classroom with tables and chairs and the other in a long corridor (1969). Distance x_i was varied 5 ways from 5 to 45 ft in the former and 9 ways from 5 to 80 ft in the latter. Each S gave two magnitude estimations z_i to each x_i without a fixed modulus and individual power functions were fitted (16 S's and 18 S's). The distribution of individual values of slope β was the same in the two experiments. Its average was 1.21. In Fig.4.15, two straight lines having the slope of 0.87 and 1.21 are shown approximately over the range of x in their experiments. The scale of ordinate $d(x)$ is with an arbitrary unit.

In magnitude estimation, stimulus is presented one by one. Differing from the method of equal-appearing intervals, no adjustment of stimulus position by the S is necessary. However, two conditions are important. One is that successive presentations should be done quickly, because all assessments are relative. The other is that there should not be any after-effect of one stimulus presentation to the next. In the experiment of Da Silva and Das Santos, the experimenter quickly moved the target from one position to the next and the S was prevented from seeing this move.

In the experiments III and IV of Galanter and Galanter, it would have taken some time for the aircraft to come to the next position after it flew over one position. Nothing was mentioned as to the average interval of time between successive magnitude estimations.

In addition to magnitude estimation, we have another scaling procedure that is based on ratio judgments on δ. A standard target is presented at x_i and the position y_i of the variable target is adjusted until the following condition is satisfied.

$$\delta(y_i) = K \delta(x_i), \quad K = 1/2 \text{ or } 1/3, \text{ etc.}$$

This is called the *method of fractionation*. From these data (y_i, x_i), $i = 1$, 2, ..., N, it is easy to obtain $d(x)$ as a power function (2.4.8) with an arbitrary unit. By this procedure, Purdy and Gibson (1955), Da Silva showed in outdoor experiments that β was close to or less than 1.0 (1982, 1983). Cook (1978) obtained a similar result.

4.2.3. *Discussion on the Form of d(x)*

Let us consider the general case $y(x)$ first. According to the kind of scales of x and y, the possible functional relation between these two variables is restricted. Herein, the discussion is limited to two kinds of scale only, *ratio scale* and *interval scale* (Stevens, 1951; Krantz et al., 1971) and to cases in which only one independent variable x is involved.

A scale whose permissible transformation is change of unit only is called a ratio scale. By the change of unit, the same magnitude can be expressed as ϕ or ϕ' where

$$\phi' = k \phi, \quad k > 0.$$

The name "ratio scale" comes from the fact that, in order to define the ratio between a and b, these must be expressed in terms of a scale of this kind. Otherwise, $\phi(a) / \phi(b) \neq \phi'(a)/ \phi'(b)$. Most physical measurements are given in ratio scales.

A scale, in which changes of origin as well as of unit are permissible, is called an interval scale. Celsius temperature and Fahrenheit temperature are defined with different units and different zero points. In theoretical equations, temperature is given in the ratio scale Kelvin. The permissible transformation for an interval scale ϕ is

$$\phi' = k\phi + C, \quad k > 0$$

The difference between a and b measured in ϕ and the difference measured in ϕ' are proportional. However, the invariance of ratio, $\phi(a) / \phi(b) = \phi'(a) / \phi'(b)$, does not hold.

The possible functional form of $y(x)$ is constrained by the scale kinds of both variables (Luce, 1959). Consider the cases in which x is a ratio scale,

(I) If y is a ratio scale, then $y(x)$ is a power function

$$y(x) = \alpha x^{\beta}, \text{ where } \beta \text{ is independent of the units of } x \text{ and } y \quad (4.2.9)$$

(II) If y is an interval scale, then there are two possible forms

$$y(x) = \alpha \log x + C, \text{ where } \alpha \text{ is independent of the unit of } x$$
$$(4.2.10)$$
$$y(x) = \alpha x^{\beta} + C, \text{ where } \beta \text{ is independent of the units of } x \text{ and } y$$
$$C \text{ is independent of the unit of } x$$

Suppose that $y(x)$ is transformed to $y(kx)$ due to change of x to kx. Eq.(4.2.9) is the solution of a functional equation, $y(kx) = K(k)y(x)$, and Eq.(4.2.10) is the solution of a functional equation, $y(kx) = K(k) y(x) + C(k)$.

(III) If the independent variable x is dedimensionalized by a constant $a(\neq 0)$, x/a does not have unit. Hence, the above stated constraints on the functional form are not applied to $y(x/a)$.

Let us consider the form of distance function $d(x)$ from this point of view. The independent variable x for physical radial distance is a ratio scale. Consider the case that the dependent variable $d(x)$ is a scale satisfying Eq.(4.2.7). Though I hesitate to call it a ratio scale for the latent variable $\delta(x)$ because the scale character of δ is not clear, it is still true that the permissible transformation to $d(x)$ is change of unit only. In this sense, it is understandable from (I) that all the results in Fig.4.15 are of the form of Eq.(4.2.9). The curvilinearity at the lower end in Fig.4.15 suggests that there exists a minimum distance $x_0(>0)$ from which the physical distance becomes perceptually meaningful. Then, $(x - x_0)$ is a ratio scale and $d(x) = \alpha(x - x_0)^\beta$. When x is large, the effect of x_0 is negligible but when x becomes closer to x_0, the plotting of log $d(x)$ against log x_0 starts to curve downward.

As discussed with regard to ratio assessments r in Section 3.2.1, Eq.(4.2.7) holds even when it is not δ but δ^m that is proportional to d. Then, suppose that $d(x)^{1/m} = ax^b$ holds,

$$d(x) = (a^m) x^{(mb)}$$

Hence, in Eq.(4.2.8), $\alpha = a^m$ and $\beta = mb$. Namely, the exponent β is determined by how ratio assessments are related to δ. It is well known that, when Eq.(4.2.8) is fitted to individual data, there are fairly large individual differences in the value of exponent β. This may be due to the involvement of m. It is likely that individual differences in b are not large but those in m are large. The value of β tends to change according to the dynamic range of x_i in the experiment (*e.g.*, Da Silva, 1982), the smaller the range the larger the value. This tendency may be also ascribed to the change of m, rather than that of b.

Suppose that d in Sec. 4.2.1 can be regarded as an interval scale in the sense that differences $[d(x_i) - d(x_j)]$ of the same size Δ are based on the same increment $\varepsilon - [\delta(x_i) - \delta(x_j)]$. Then, according to (II), $d(x)$ must be of the form of either one in Eq.(4.2.10). The curves shown in Fig.4.12 are not of logarithmic form. If we think of a curve running through the four curves, Kuroda I, II, Gilinsky, and Teichner *et al.*, it can be approximated by a power form of Eq.(4.2.10) with an appropriate value of C. If C is about 10 sm, then the slope β is about 0.3. This may not be a meaningful representation, however. The series of equal-appearing intervals in the experiments of Gilinsky, Harway, and Kuroda were constructed right from the foot of the S. In contrast to equisection experiments in other modalities, the origin of these curves in Fig. 4.12 is not arbitrary. The general trend of those curves can be also represented by Eq.(4.2.2) (broken curve labeled Gilinsky in Fig.4.12). The equation can be written in the following form.

$$\frac{d}{D} = \left(\frac{x}{M+x}\right)$$

In this equation, d is dedimensionalized by a parameter D on the left side, and x is dedimensionalized in the parenthesis on the right side. According to (III), this is a case in which we cannot specify the functional form between d and x from their scale natures.

We have another case in which the functional form is not constrained. The Luneburg's mapping function $g(\gamma)$ in Eq.(2.3.1) relates the Euclidean distance ρ_0 in EM to the convergence angle γ (Sec.2.3.1). These two variables are both ratio scales. Nevertheless, $g(\gamma)$ is not a power function. This is not a violation of (I), because the independent variable in Eq.(2.3.1) is not γ but $(\sigma\gamma)$ and γ is dedimensionalized by σ^{-1}.

In general, when two scales of sensory intensity for the same stimulus of intensity s are constructed, $u(s)$ based upon judgments on differences and $v(s)$ based upon assessments of subjective ratios, these two are not linearly related. When $v(s)$ is plotted against $u(s)$, the trend is an accelerated curve (convex downward) in many modalities. In Figs. 4.12, all scales d based on difference judgments are concave downward in log-log coordinates and in Fig.4.15 all scales d based on ratio assessments are almost linear in log-log coordinates, which suggests that the relationship between these two scales of δ is not an exception. Furthermore, it was found that, when the maximum dynamic ranges of brightness, loudness, and sweetness in taste are converted to the same size, the curvilinearities between $v(s)$ and $u(s)$ become almost the same in these three modalities (Indow and Ida, 1977). In order to see if the same curvilinearity appears between two scales of δ, we need to know max d in each scale.

If we select N targets $\{Q_i\}$ in front of the body Q_0, all in the direction of x-axis, sufficiently extended within a narrow range of ϕ_i or θ_i, we can ask the S to assign numerical values $r_{i,j\alpha}$ in the form of Eq.(3.2.1), and define $\mathbf{D} = (d_{ij})$ where $i, j = 0, 1, 2, \ldots, N$. Then we can apply the method used in the multidimensional construction of star configuration (Sec.4.1.3). Differing from stars, we have information on physical distances $x_0 = 0$, $x_i > 0$, $i = 1, 2, \ldots, N$, and we can relate the obtained scale d to x_i. Namely, the result $d(x_i)$ will be more informative than in Sec.4.1.3.

4.3. Perceived Spatial Layouts under Full Cue Conditions

Haber (1985) wrote "Most current perceptual theories do not consider perceived layout, preferring to focus on the much narrower concern of perceived radial distance of objects from the observer, ignoring concern for the relationships among the objects of the scene, and among the observer in relation to the objects." In Sec.4.2, the methodology to scale perceived radial distances was discussed. In this section, three experiments concerning perceived spatial layouts will be reviewed from the same point of view. Haber extensively cited an experiment by Toye (1986).

4.3.1. Three Experiments

The basic form of Toye's experiment was as follows. Thirteen white metal stakes, Q_i, were semi-randomly placed within an area spanning a diameter of about 70 ft in the center of flat grassy field, 250 ft on a side. A building and trees were visible beyond the field. The S sat at the position of stake "0" and assessed inter-stake distances e_{ij} to give scaled data d_{ij} by moving the head freely. A picture of this layout $\{Q_i\}$, $i = 0, 1, 2, ..., 12$, is shown in Fig.3 of the Haber's article. The viewing position, the stake "0", was altered in two ways so as to see the configuration from one side "A" and from the other side "B" that is perpendicular to A. In the present context, we do not need to make distinction between these two viewing directions because the coordinate axes, x and y, were always defined with regard to stake "0". The assessment was made in three ways in a fixed order. Eight S's repeated each kind of judgment twice. Repeated results were treated separately and hence the total number of data matrices $\mathbf{D} = (d_{ij})$ was $48 = 3 \times 8 \times 2$. From each $\mathbf{D}$, 13×13, a configuration $\{P_i\}$ was constructed in E^2, 2-D Euclidean plane, by a nonmetric multidimensional scaling procedure (MDS). This MDS is different from EMMDS (Sec.3.2.1) in two ways. First, $\{P_i\}$ is directly constructed in E^2 under the assumption that it represents the perceptual configuration $\{P_{Vi}\}$ in VS of the S. It is equivalent to assume that $\{P_{Vi}\}$ is of Euclidean structure. Second, only a monotonic relationship, not necessarily proportionality, is assumed between data d_{ij} and interpoint Euclidean distances $\hat{d}_{ij}$ in $\{P_i\}$. In other words, $\{P_i\}$ is defined on the basis of order information among d's, not of numerical information. The unit by which $\hat{d}_{ij}$ are defined is arbitrary. The input matrix **D** was obtained in three ways.

I. Relative judgment. As to each of 286 possible sets of 3 Q's, (e_{ij}, e_{jk}, e_{ik}), the S indicated which one of the three distances was the largest. The number of times that e_{ij} was endorsed as being the largest was defined as d_{ij}.

II. Absolute judgment. To each of 72 possible pairs, (Q_i, Q_j), the S assigned a value d_{ij}(sft) that represents δ_{ij} in terms of the subjective length of a foot. This is the same idea as in the Gillinsky's definition of d_0 (Sec.4.2.1).

III. Map drawing. The S drew a map of the scene on a 20-cm square sheet of paper. Distances d_{ij} in this map were measured.

Toye described in detail that $\{P_i\}$'s constructed in E^2 were highly consistent with original data **D** and highly stable across repetitions, and then compared each of 48 $\{P_i\}$'s with $\{Q_i\}$. Because the orientation of $\{P_i\}$ and its coordinate units are different from those of $\{Q_i\}$, an appropriate rotation and adjustment of units are necessary for this comparison. His results are summarized in the framework of the current context. As in Figs.1.1 and 1.3, the coordinates of Q are denoted as (x, y). Toye called x as radial extent and y as horizontal extent. Denote by ξ_V, η_V two orthogonal directions of $\{P_i\}$ that is matched with $\{Q_i\}$ in orientation and size. It was found that $\eta_V = y$, and $\xi_V = cx$, $c < 1.0$. Perceptual distances in the radial direction are considerably contracted in appearance compared with those in the horizontal direction. If I understand correctly, the mean value of c is about 0.85 in the case of $\{P_i\}$ based on d_{ij}(sft) in II. In order to be compared with $\{Q_i\}$ directly, $\{P_i\}$ must be stretched by 1/c in the radial extent. Let us denote this stretched $\{P_i\}$ as $\{P'_i\}$. Toye showed in his Fig.5 $\{Q_i\}$ and an example each from $\{P'_i\}_\alpha$, α = I, II, III, according to the kinds of judgment by which d's were defined. In this Figure, $\{P'_i\}_{II}$ looks most similar and $\{P'_i\}_{III}$ least similar to $\{Q_i\}$. According to Haber, in general, $\{P'_i\}_I$ and $\{P'_i\}_{II}$ are about the same whereas $\{P'_i\}_{III}$ is slightly but significantly less accurate in reproducing $\{Q_i\}$, and he wrote "I was surprised that the drawn maps resembled the actual scene less accurately than the constructions based on the two distance estimates". Neither Toye nor

Haber made explicit the form of relation between data d_{ij} and interpoint distances $\hat{d}_{ij}$ in $\{P'_i\}_\alpha$. It must be closer to proportionality in $\{P'_i\}_{II}$ than in $\{P'_i\}_I$. In $\{P_i\}_{III}$, we can measure angles in addition to distances. This information was not used, however. This study is based on the assumption that the perceptual configuration $\{P_{Vi}\}$ is of Euclidean structure and it is the replica of the physical spatial layout $\{Q_i\}$ with "shortening of radial distances".

Levin and Haber (1993) repeated the experiment of Toye. Under the similar condition, a configuration of 13 stakes, $\{Q_i\}$, was observed by eight S's with no previous experience in psychophysical experiments. As in the Toye's experiment, the S sat at a stake "0" and made magnitude estimation d_{ij} with the unit of sft to give $\mathbf{D} = (d_{ij})$. The viewing position, stake "0", was altered in two ways to view the configuration from one side "A" and from the other side "B" that is perpendicular to A. In this experiment, the change of viewing direction played an important role. In contrast to the Toye's experiment, Q_i were located to meet some criteria (Fig.1 in their article). For instance, some pairs of Q's were parallel to or perpendicular to the line of sight according to the viewing direction. First, they plotted 78 interpoint distances in Fig.2 of their article, mean value (2 repetitions of 8 S's) d(sft) vs. physical distance e(ft). The suffixes, i, j, etc., are omitted. Points scattered very closely along a straight line passing through the origin with the slope of 1.11. Then, using the following regression equation, they tried to reduce the scatter and/or to account for the "overestimation of e's by about 10 %". Let us denote by $\Delta\phi$ the angle spanned by two vectors from "0" to Q_i and Q_j, $\Delta\phi = |\phi_i - \phi_j|$, where ϕ is the lateral angle (Fig.1.3) and $\Delta\phi$ ranged from 0° to 82°.

$$d(\text{sft}) \Leftrightarrow \hat{d} = \alpha_e\, e + \alpha_\phi\, \Delta\phi \qquad (4.3.1)$$

Then, $\alpha_e = 1.108$, $\alpha_\phi = 0.164$ and the scatter is reduced. They reanalyzed the data of Toye in the same way and found $\alpha_e = 0.865$, $\alpha_\phi = 0.175$ and came to the conclusion opposite that of Toye. The change of perceived length of e according to its orientation is not due to the contraction of distance in which $\Delta\phi = 0$ but to the expansion of distance that includes the additional contribution of the $\Delta\phi$. They call it "visual angle effect". In the second experiment, they showed that this "overestimation" due to

140 *Global Structure of Visual Space*

$\Delta\phi$ was independent of the way by which $\Delta\phi$ came about, by the orientation or by the viewing distance, *etc*. As they admit at the end of the article, the "overestimation of *e*'s" presupposes that physical foot (ft) and subjective foot (sft) are commensurable and equivalent. They wrote "if all S's had an internalized metric that was on the average 10% smaller than the physical unit foot, then, on the basis of the present data, horizontal dimension of space would be accurately estimated and radial dimensions would be underestimated". It seems to me meaningless to compare sft and ft directly. That is the reason I tried to keep the two separate in Eqs.(4.2.1) and (4.2.2). However, the expression that I wrote before, "if the unit of η_V is made equal to that of *y*, $\eta_V = y$, then ξ_V is shorter than *x*, *i.e.*, $\xi_V = cx$, $c < 1.0$", holds for the results of the experiments of Toye as well as for the results of Levin and Haber.

A similar experiment was performed by Wagner (1985). The spatial layout $\{Q_i\}$ consisted of 10 stakes uniformly distributed within an area, 40 m deep and 40 m wide, in a large open field under full cue conditions. No stakes could deviate in terms of ϕ by more than $\pm 45°$ and no two stakes could be closer than 5 m from each other. This $\{Q_i\}$, $i = 1, 2, ...,$ 10, was observed from three different positions, the nearest being the middle point of the near edge of the stake field and the furthest being 40 m apart from that middle point. From each position, five S's made four kinds of estimation (magnitude estimation, category estimation, map drawing, and perceptual matching of angles) on physical inter-stake distances e_{ij}, and angles $\omega_{i,j,k}$, as well as areas of triples of stakes in $\{Q_i\}$. Two S's were familiar with these judgments and three S's were naive. The S was allowed to move the head freely during the estimation. In all the cases, Wagner analyzed the data under the assumption that the estimated value is a power function of the corresponding physical variable. In this context, only the results of magnitude estimation on distance e_{ij} and angle $\omega_{i,j,k}$ are described. A triangle of three stakes was presented to the right of the S at a distance of 10 m to define the modulus for the magnitude estimation. A value of 100 was assigned to a side of this triangle for magnitude estimation on e_{ij} and to an angle of this triangle for magnitude estimation on $\omega_{i,j,k}$. Again, the suffixes, *i*, *j*, are omitted. As shown in Fig.4.16, let us denote by $\hat{d}$ the expected value of magnitude estimation *d* on distant *e* (Dist. I) and by $\hat{a}$ the expected value of magnitude estimation a on angle ω (Ang. I). Then, $\hat{d}$ and $\hat{a}$ are assumed to be power functions of e and of ω respectively.

$$d \Leftrightarrow \hat{d}(e) = \alpha e^{\beta}, \quad \alpha > 0, \ \beta > 0 \qquad (4.3.2)$$
$$a \Leftrightarrow \hat{a}(\omega) = \alpha \omega^{\beta}, \quad \alpha > 0, \ \beta > 0 \qquad (4.3.3)$$

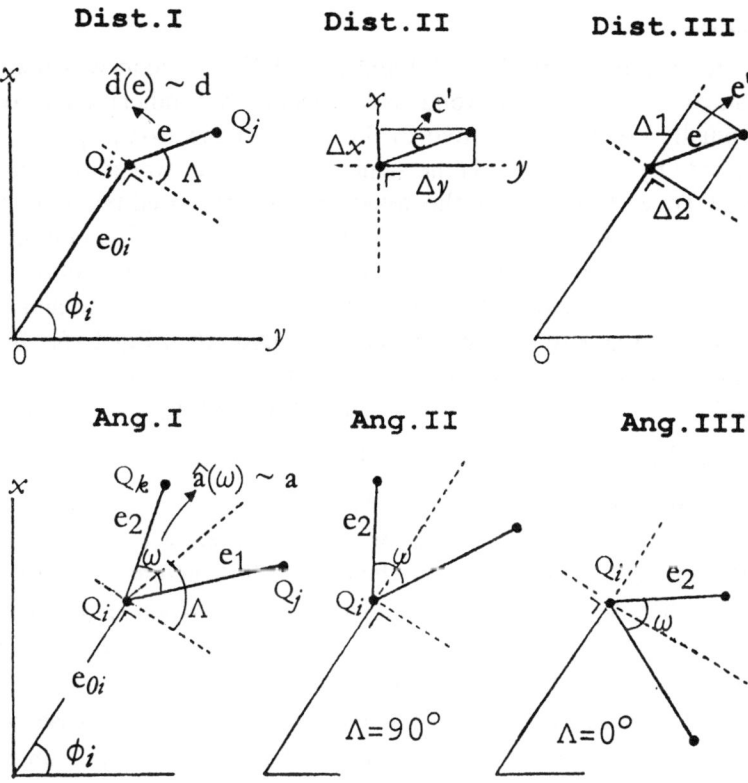

Fig.4.16 Judgments on distances and angles in Wagner's experiment

A point Q in the field is localized as either (x, y) or (e_0, ϕ). With the same Q, these coordinate values change according to the observing position of the S. When e in Dist.I is at issue, Q_j is localized with regard to Q_i by e and the orientation angle Λ. When ω in Ang I is at issue, Q_j and Q_k are localized with regard to Q_i by two legs of angle e_1, e_2 and the orientation angle Λ defined with regard to the bisecting direction of ω. These coordinates are independent of the observing position. Fits of $\hat{d}(e)$ to data d or $\hat{a}(\omega)$ to data a are shown in Table 1 of his article where data includes all values of e_0, ϕ, and Λ. Goodness of fit is given by r^2 between log d and log $\hat{d}(e)$ or r^2 between log a and log $\hat{a}(\omega)$. The fit was made to individual data as well as to the composite data of 5 S's. Herein only the results of $\hat{d}(e)$ and $\hat{a}(\omega)$ for the composite data will be discussed.

In the case of $\hat{d}(e)$, $\beta = 0.99$ and $r^2 = 0.77$. A trend was noticed: "distances seen in depth (one stake behind the other) tend to be perceptually contracted relative to the those in the frontal plane (two stakes side by side)". This is the same tendency that was noticed in the study of Toye. Coping with this trend, Wagner proposed two models to replace e in Eq.(4.3.1) by e' and e''. One is called "affine contraction model" and the other "vector contraction model". In the former, the contraction occurs along the x-axis and e' is defined by Δy and $c\Delta x$, where $c < 1.0$ and Δx, Δy are explained in Dist. II in Fig.4.16. In the latter, the contraction occurs in the direction of depth for the S and e'' is defined by $\Delta 2$ and $c\Delta 1$, where $c < 1.0$ and $\Delta 1$ and $\Delta 2$ are explained in Dist. III. In the case of $\hat{d}(e')$, $\beta = 0.97$, $c = 0.49$ and $r^2 = 0.74$. In the case of $\hat{d}(e'')$, $\beta = 0.93$, $c = 0.46$ and $r^2 = 0.75$.

The same angle ω is perceived with different sizes according to the orientation angle Λ. It is largest when $\Lambda = \pm 90°$ such as in Ang.II and smallest when $\Lambda = 0°$ and $180°$ such as in Ang.III. When $\hat{a}(\omega)$ is fitted to data a of all values of e_0, ϕ, and Λ, $\beta = 0.81$, $r^2 = 0.82$. Wagner considered angles under two contraction models to have $\hat{a}(\omega')$ and $\hat{a}(\omega'')$. The value of β was of the same level (0.81) in either case, but r^2 became slightly smaller. Always, $\beta < 1.0$ and $\hat{a}(\omega's)$ are functions concave downwards. No attempt was tried to construct the perceptual layout $\{P_{Vi}\}$ on the basis of $\hat{d}(e's)$ and $\hat{a}(\omega')$ or to relate $\hat{d}(e's)$ and $\hat{a}(\omega')$ to distances and angles in the drawn map.

To relate d directly to e, e', e'', Euclidean distances in the physical space X^2, is equivalent to assuming that the perceived configuration $\{P_{Vi}\}$ is of Euclidean structure. He tested a possibility that d is a power function of Minkowski's power metric composed of $c\Delta x$ and Δy. This metric includes Euclidean metric as a special case, and the results showed that this special case turned out to be the case. Wagner also discussed and discarded two non-Euclidean models of VS, the spherical geometry model of Reid and the hyperbolic geometry model of Luneburg. The former is a model proposed in 1764 in which VS is assumed to be "equivalent to the proximal stimulus at the retina". It will not be necessary here to go into it. When discussing the model of Luneburg, he did not distinguish X^2 in which Q's are presented and EM^2 in which hyperbolic VS^2 is to be represented. This is not a proper way to test the model.

4.3.2. General Discussion

In all the studies so far reviewed in Secs.4.2 and 4.3, scaled distances d_0 and d are directly related to the physical counterparts e_0 and e, which is equivalent to assume that the perceived configuration $\{P_{Vi}\}$ is of Euclidean structure. One way to approach the geometrical property of $\{P_{Vi}\}$ is to follow the procedure described in Sec.4.1.3.

By this procedure, we can construct a configuration of points $\{P_i\}$ in EM^2 that represents the perceived layout $\{P_{Vi}\}$ in VS^2. The value of K and the mapping functions need not be known in advance. It is necessary to have data matrix $\mathbf{D} = \{d_{ij}\}$ in which all d's can be assumed to be related to perceptual distances δ by a power function and radial distances d_{0i} from the self must be included. The data matrix $\mathbf{D}$ of the experiment of Wagner cannot be used for this purpose because d_{0i} were not included. The data $\mathbf{D}$ of Toye and $\mathbf{D}$ of Levin and Haber can be used because d's from the stake Q_0, at which the S sat, are regarded as d_{0i}. Furthermore, we need to assume that d obtained by the direct magnitude estimation is related to δ in the way as mentioned above. As stated at the end of Sec.4.2.3, however, magnitude estimation tends to introduce a bias in the data. The ratio judgments described in Sec.3.2.1 is a procedure to minimize the possible bias. However, compared with the direct magnitude estimation, it takes much more time to obtain a $\mathbf{D}$ by this method. It was pointed out in Sec.4.1.3 that constructing $\{P_i\}$ in EM^3 from $\mathbf{D}$ was not powerful enough to distinguish three geometries. The same situation may occur with a spatial layout $\{Q_i\}$ as used in these two experiments. The best way to avoid this possibility is to use a layout

$\{Q_i\}$ that yields $\{P_{Vi}\}$ having some geometrical property. In the wide field under full cue conditions, such a layout $\{Q_i\}$ can be constructed in which Q's are located by the S so as to meet a prescribed geometrical property. In the experiment of Levin and Haber, Q's were located to meet some physical criteria. We can ask the S to locate Q's to meet some perceptual criteria. For instance, some series of Q's are perceptually collinear and a series of Q's are perceptually parallel with or orthogonal to another collinear series of Q's, *etc*. If **D** is obtained by the ratio assessment (Sec.3.2.1) with this $\{Q_i\}$ including d_{0i} and $\{P_i\}$ is constructed in EM^2 from this **D**, the result would be more informative. Interpoint distances $\hat{d}_{ij}$ and the coordinates (ξ, η) of $\{P_i\}$ in EM^2 are related respectively to latent perceived distances δ_{ij} and to the coordinated (x, y) of $\{Q_i\}$ in X^2. By plotting ξ against x and η against y, it will become automatically clear which one is contracted.

It is not necessary to constrain the configurations to be 2-dimensional. Using stakes of different heights, we can have a 3-D configuration $\{Q_i\}$ in X^3. For instance, some sets of Q's appear to be frontoparallel at a certain distance and other sets of Q's appear to be frontoparallel at another distance, *etc*. If we ask the S's to draw a 3-D trompe l'oeil in addition to assessment of e's, the information from this map can be used as the initial configuration to obtain a $\{P_{Vi}\}$ in VS^3. Should the hypothesis stated in Sec.3.2.3 be tested, in comparing $\hat{d}_{ij}$ with data d_{ij}, we can define $\hat{d}_{ij}$ of $K = 0$ for (P_i, P_j) in the same HP and $\hat{d}_{ij}$ of $K < 0$ for (P_i, P_j) when P_i and P_j are in two different HP's. In contrast to the approaches of Toye and of Wagner, it is not necessary in this approach to assume that $\{P_{Vi}\}$ is a veridical replica of $\{Q_i\}$.

Cutting and Vichton (1995) divided VS into three zones according to what cue is effective to produce difference in radial distance δ_0 from the self in VS. In the personal space ($e_0 < 2$ m), convergence angle γ is effective in determining δ_0 (Fig.1.3). In the action space (2 m $< e_0 < 30$ m), the binocular parallax and motion perspective are the main factors to determine δ_0. In the vista space ($e_0 > 30$m), δ_0 is not determined by these egocentric cues. They mentioned as the non-egocentric factors to determine δ_0, occlusion, relative size of familiar objects, aerial perspective, and relative density of surface, *etc*. Heelan (1983) distinguished near zone and distant zone in VS. It is evident that VS is divided into zones and different mapping functions hold according to zone. However, it is not clear whether VS exhibits different geometrical structures according to zone. We do not see any discontinuity between personal space and action space even if the underlying cues are not the same.

In the neighborhood of the self, VS under natural conditions is veridical to X. Otherwise, we cannot manipulate physical objects or move around in X (VS.4 in Section 1.1.1). If the point-to-point correspondence between VS and X is considered, it must be very complicated in the neighborhood space under natural conditions. I am not sure as to whether $\{P_{Vi}\}$ discussed in Sec.4.3 is in this neighborhood space or not. On the other hand, VS's discussed in Secs.4.1 and 4.2 are beyond this neighborhood space.

4.3.3. *Regular Triangles with S at the Barycenter*

A very unique experiment was conducted in the daylight in a flat terrain with weeds covering the ground (Koenderink and van Doorn 1998; Koenderink, *et al*, 2000). Some trees and buildings were visible at the far distance. The S stood at the barycenter of a regular triangle defined by the three Q's at the eye-height. Fig.4.17I is a schematic illustration of two out of six regular triangles of different sizes. A filled dot indicates the positions where a target T (a bright orange sphere subtending 20 ± 4 min of arc) was presented and an unfilled dot the position at which the pointer was placed. The pointer is an orange arrow sticking out 50 cm from a white cube (25 cm edge length) at both sides. Fig.4.17II illustrates the task of S in a stimulus layout (an isosceles) corresponding to a part of a regular triangle in I. The experiment was performed with each of the six regular triangles. The side lengths e are shown in the Figure. The S rotated pointer P via remote control to point at target T. This is called "exocentric pointing" in contrast to "egocentric pointing" in which the S directly points at a target as in directing a gun. In each stimulus layout, pointer P and target T spanned $120°$, and the S was allowed, insofar as the eye height was maintained constant, to move and turn the head, to twist at the hip, and to change feet position. It was written, "scatter in the settings (by the method of adjustment) is about $1.5°$, which appears to indicate remarkably accurate performance". It

146 *Global Structure of Visual Space*

was not stated how many times the settings were repeated. Three S's participated and the results were analyzed individually.

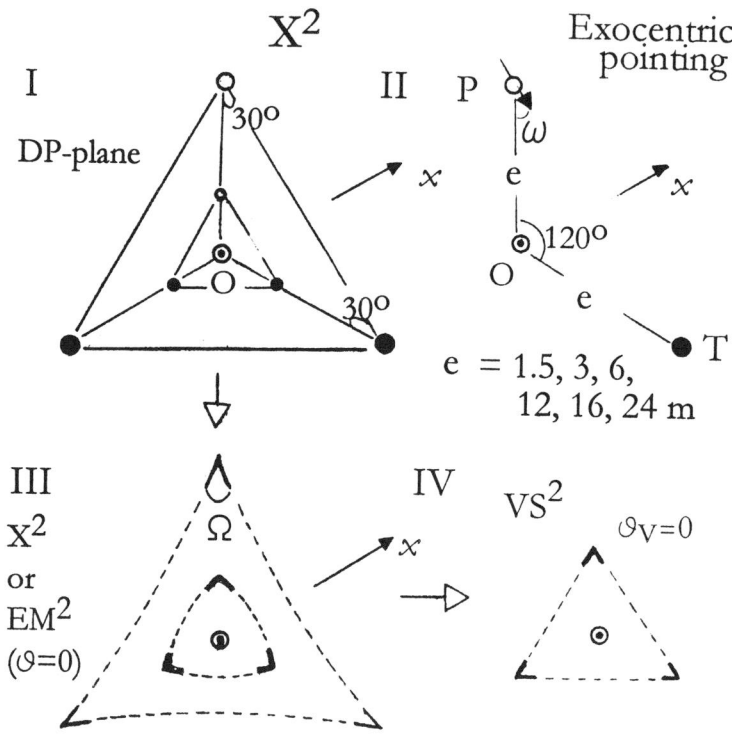

Fig.4.17 Exocentric pointing experiments by Koenderink *et al.*

Denote by ω the mean settings of an angle in II. An angle Ω of a regular triangle can be obtained from ω. If all the settings are veridical, Ω is 60°. The symmetrised result of a S in X^2 was shown in the original articles with the comment that "the results of other S's were similar". Bold segments in Fig.4.17III illustrate angles Ω of two triangles in X^2 to show the tendency that $\Omega > 60°$ in smaller triangles and $\Omega < 60°$ in larger triangles. The length e of the regular triangle in which Ω is about 60° is 3 m for two S's and 12m for one S. The investigators designed and performed this experiment with the purpose to approach the basic geometrical structure of VS without using any metric information. Let us begin with discussing their results in the framework of this book.

When the S stands at the barycenter of a triangular configuration of three stimulus points $\{Q_i\}$ in Fig.4.17I, the perceived layout $\{P_{Vi}\}$ will be as shown in IV. The exocentric pointing means that the pointer is set along the imaginary side indicated by the broken line. The change of Ω according to the size of regular triangle can be interpreted in two ways.

(A) If we assume that each $\{P_{Vi}\}$ is structured with its own K, the triangular pattern $\{P_i\}$ in EM^2 that represents $\{P_{Vi}\}$ changes its form according to the sign of K. In this interpretation, Fig.4.17III shows two $\{P_i\}$'s in EM^2, each representing the perceived regular triangle (IV). Since the head of S is not fixed in setting the pointer, we cannot define $\{P_i\}$ from $\{Q_i\}$ in X^2 by Luneburg's mapping functions. If we assume that the mapping from X^2 to EM^2 under this condition is conformal, then K > 0 in the small triangle and K < 0 in the large triangle.

(B) The setting in II can be regarded as an operation to define the frontoparallel curve connecting P and T for the S facing toward the x-direction. In this case, Fig.4.17III represents stimulus patterns in X^2, each yielding the perceived regular triangle in VS^2 (IV). Then, the broken curves in III are similar to the H-curves in X^2 (the horizontal DP in Fig.2.1), though the span is much larger. The H-curves in Fig.2.1 are obtained with a fixed value of K. As explained in Sec.2.2.3, changing of H-curves from concavity to convexity is not related to the value and sign of K. It is due to the mapping relationship between EM and X. In the same token, it may be possible to define a set of appropriate mapping functions, not necessarily of the form of the Luneburg's functions, that account for the systematic change of angles Ω under the assumption that K is constant for all regular triangles at issue.

Koenderink and collaborators regard their results to be against the Luneburg's postulate that VS is homogeneous and is of constant K. They are concerned with the homogeneity of the entire VS in which each of the six regular triangles is perceived. In their approach, EM and mapping functions between EM and X are not considered, and hence the interpretation (B) was not discussed. Under this interpretation, their results are not necessarily inconsistent with the postulate of Luneburg.

According to the interpretation (A), their results falsify the homogeneity of the entire VS but do not contradict the assumption that K is constant in each $\{P_{Vi}\}$. To assume the homogeneity for the entire VS means that K is the property of VS no matter which regular triangle is being perceived therein. What is emphasized in this book is that K is the structural property of a $\{P_{Vi}\}$ being perceived. Then, their results indicate that K systematically changes as a function of the size of $\{P_{Vi}\}$ and hence the distance from the self. Using the relationship that [angular excess (the sum of Ω's $- 180°$) times the area] is equal to K, they showed a figure depicting the change of K in X^2. According to the interpretation (B), there is the possibility that all triangular patterns $\{P_i\}$'s in EM^2 are of the same form with a given value of K, but their projected forms in X^2 are different as observed in the experiment.

5. Related Experiments and Theoretical Considerations

5.1. Spatial Layouts in Frameless Visual Space

Concerning the Luneburg model of VS, there are a large number of articles, some are experimental and some are theoretical. As to experimental studies, only P- and D- alleys and H-curves were dealt with in Chapter 2. In Sec.5.1.1, some of other experiments will be reviewed. As in most alley experiments, all were performed with $\{Q_i\}$ of small light points in a dark room. The review is not intended to be exhaustive. In the subsequent sections will be reviewed experiments and theoretical considerations on mapping functions and the basic postulate that VS is a Riemannian space of constant curvature R.

5.1.1. *An Experiment with Circles*

By asking the S to construct a $\{Q_i\}$ that appears as a circle in VS, Hagino and Yoshioka (1976) tried to estimate the individual constants K and σ. Both assumptions in the Luneburg model, VS being a Riemannian space of constant negative K and the mapping functions, Eqs. (2.3.1, 2), were taken together.

In addition to a point Q_F in the center, 16 Q_i were simultaneously presented in a dark room, $i = 0, 1, 2, \ldots, 15$, where Q_F and Q_0 were fixed on the *x*-axis. The distance (Q_F, Q_0) is denoted as R (Fig.5.1). All light points Q's were on a horizontal plane slightly below the eye-level to avoid occlusion. As the elevation angle θ of the farthest point Q_0 was only about 40 min of arc, all were regarded as being on the DP($\theta = 0$).

Let us denote by δ_R the perceptual distance generated by the physical distance R between Q_F and Q_0. The positions of other Q_i were moved in the direction of ϕ'_i until the S judged that the distance from Q_F to Q_i appeared equal to δ_R (method of limits). Once the adjustments of all Q's were made, 16 Q's appeared to be on a circle of radius δ_R around P_{VF}. The S was allowed to readjust any Q_i whenever it is felt necessary. These adjustments were repeated four times and the mean positions of $Q_i(x_i, y_i)$ were obtained with each S. By taking the mean of $Q_i(x_i, \pm y_i)$, they defined $\{Q_i\}$ that is symmetric with regard to the x-axis. As shown in Fig.5.1, the position of Q_0, x_0, was varied in three ways with R = 50cm (small circles in the close range) and in four ways with R = 100cm (large circles at the far distance). In this way, seven $\{Q_i(\gamma_i, \phi_i)\}_{FR}$ were obtained with each of five S's. The positions of Q_F and $Q(\gamma, \phi)$ in EM^2 are shown as $P_F(\rho_{0F}, 0)$ and $P(\rho_0, \varphi)$ (the right side in Fig.5.1). In a $\{Q_i(\gamma_i, \phi_i)\}_{FR}$, ρ_{0F} corresponding to $(x_0 - R)$ is fixed and each ρ_0 is adjusted so that ρ_{FP} gives rise to the fixed δ_R.

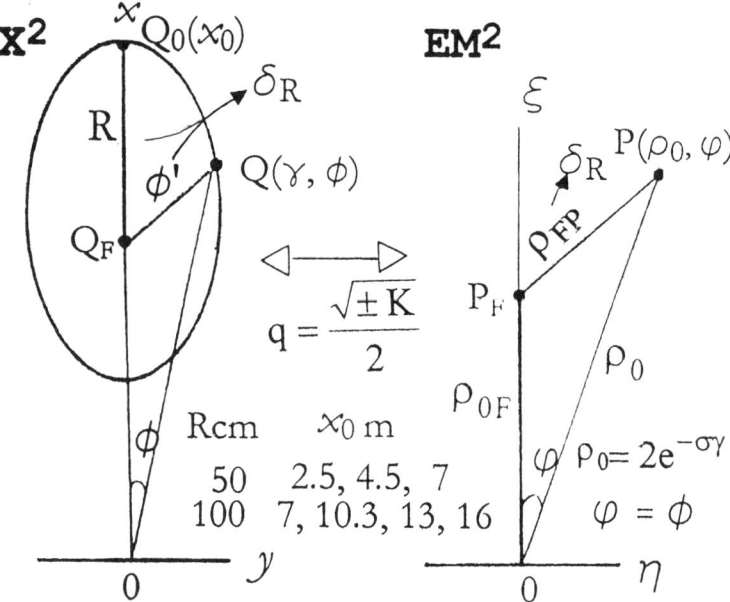

Fig. 5.1 Spatial layout on the horizontal DP to yield a circle in VS

If $K < 0$, a circle in VS is represented in EM^2 by the locus of $P(\rho_0, \varphi)$ that satisfies the following conditions. From Eq.(2.2.1),

$$[G_F \sinh^{-1} q\delta_R] = \frac{q\rho_{FP}}{G_P} \text{ is constant}$$

where

$$G_F = \sqrt{1-(q\rho_{0F})^2}, \quad G_P = \sqrt{1-(q\rho_0)^2}, \quad q = \frac{\sqrt{-K}}{2}$$

$$(q\rho_{FP})^2 = (q\rho_{0F})^2 + (q\rho_0)^2 - 2(q\rho_{0F})(q\rho_0)\cos\varphi$$

Instead of estimating one set $(q, \sigma)_{FR}$ on the basis of data $\{Q(\gamma_i, \phi_i)\}_{FR}$, $i = 1, 2, \ldots, 15$, Hagino and Yoshioka obtained $(q, \sigma)_{FR}$ separately in each direction ϕ_i. As an example, the whole results $(q, \sigma | \phi_i)_{FR}$ of a S are given in Table 2 of their article. All K's are negative and less than -1.0. As to other S's, only $(q, \sigma | \phi_i = 90°)_{FR}$ are shown in Table 4 and $K < -1.0$ in all the cases. Most of these S's had participated in P-and D-alley experiments and yielded the standard results, $-0.388 < K < 0.00$, (the last row in Table 4). We cannot expect stable results with K and σ that were estimated on one data point $\{Q(\gamma_i, \phi_i)\}_{FR}$ only. The value of σ considerably changes according to ϕ_i and K is consistently less than -1.0, which is contradictory to the Luneburg model. It seems to me that the authors accepted that $K < 0$ and tried to ascribe K being outside the legitimate range of value (Eq.2.2.4) to inadequacy of the mapping functions. It may or may not be true. That all K's were found to be negative is simply because the hyperbolic equation was used. It will be of interest to try to estimate one set $(q, \sigma)_{FR}$ from the whole data $\{Q(\gamma_i, \phi_i)\}_{FR}$, $i = 0, 1, \ldots, 15$, by carefully checking if there is an appropriate value range of σ in which $-1.0 < K < 0$. We must take into account other possibilities also.

If $K > 0$, the locus of $P(\rho_0, \varphi)$ is

$$[G_F \sin^{-1} q\delta_R] = \frac{q\rho_{FP}}{G_P} \text{ is constant}$$

where

$$G_F = \sqrt{1+(q\rho_{0F})^2}, \quad G_P = \sqrt{1+(q\rho_0)^2}, \quad q = \frac{\sqrt{K}}{2}$$

$$(q\rho_{FP})^2 = (q\rho_{0F})^2 + (q\rho_0)^2 - 2(q\rho_{0F})(q\rho_0)\cos\varphi$$

If $K = 0$, it is the locus of $P(\rho_0, \varphi)$ satisfying

ρ_{FP} is constant

where $\rho_{FP}^2 = \rho_{OP}^2 + \rho_0^2 - 2\rho_{OP}\rho_0\cos\varphi$

We must compare which locus gives best fit to data $\{Q(\gamma_i, \phi_i)\}_{FR}$, $i = 0$, 1, 2, ... , 15. It may not be impossible to test the fit and to optimize the value of q on the assumption of φ being equal to ϕ alone without assuming the mapping function (Eq.2.3.1). For that purpose, to use a perceptual triangle seems to be more appropriate than to use a perceptual circle.

5.1.2. *Experiments using Triangles*

Blank (1961) designed an ingenious experiment with a triangle. Suppose a triangle (P_{VA}, P_{VC}, P_{VB}) in VS (Fig.5.2). Though the triangle in this Figure is drawn on the DP at eye-level ($\vartheta_V = 0$), it can be in any VS^2 and can be of any size and in any orientation. Denote by P_{VD}, P_{VE}, and P_{VF} midpoints on the repective sides and, for the sake of convenience, perceptual lengths are denoted as shown in Fig. 5.2. Then, as will be explained in Sec.5.1.3, the following relationships hold.

$$K = 0 \qquad d = \frac{c}{2} \qquad\qquad\qquad (5.1.1)$$

$$K < 0 \qquad \cosh 2d = \cosh c - \sinh^2\frac{a}{2}\sinh^2\frac{b}{2}\sin^2\omega_V$$

The hyperbolic cosine of X is monotonically increasing from $X = 0$ and the last term is positive. Hence,

$$c > 2d \tag{5.1.2}$$

$$K > 0 \quad \cos 2d = \cos c - \sin^2 \frac{a}{2} \sin^2 \frac{b}{2} \sin^2 \omega_V$$

The cosine of X is a monotonically decreasing function from $X = 0$ to $X = 180°$. Hence, for d and c in the moderate range of $X > 0$,

$$c < 2d \tag{5.1.3}$$

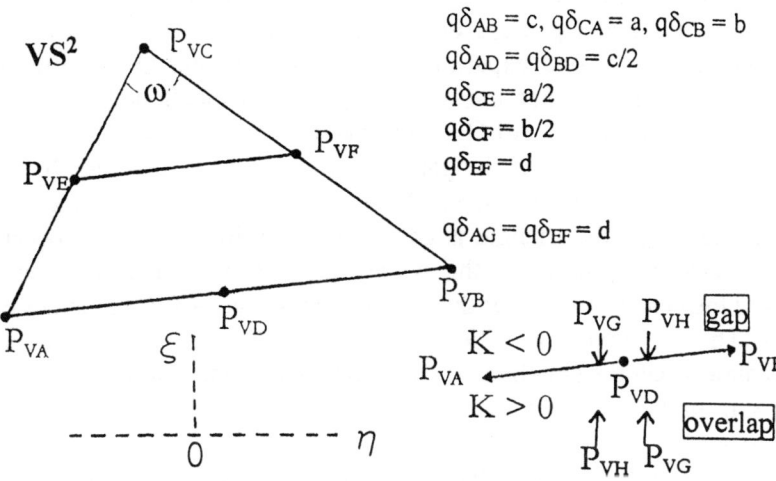

Fig.5.2 Blank's experiment (1961)

If the S determines on the side between P_{VA} and P_{VB} such two points P_{VG} and P_{VH} that satisfy,

$$q\delta_{AG} = q\delta_{EF} = d, \quad q\delta_{BH} = q\delta_{EF} = d$$

the two should conincide with P_{VD} if $K = 0$. If $K < 0$, there is a gap across P_{VD} and if $K > 0$, an overlap across P_{VD} (Fig.5.2). If we only assume that the order of collinear points P_V's in VS and the order of the correspondent stimulus points Q's in X are the same, we can experimentally determine the sign of K without using the Luneburg's mapping functions (Eqs.2.3.1, 2).

An experiment was performed with 7 S's. Three light points (Q_C, Q_A, Q_B) forming an isosceles triangle on the DP ($\theta = 0$) were kept fixed throught the experiment, Q_C (274.3, 0), Q_A(71.1, –30.5), and Q_B(71.1, 30.5) in cm. In the first experiment, the S set bisecting points Q_E, Q_F and Q_D. A trend was noticed in some S's that Q_E, Q_F shifted towards Q_C while the bisecting was being repeated. Hence, the early settings were discarded in defining (x_E, y_E) and (x_F, y_F). The results (Table I in the original article) were plotted in Fig.12 in Indow (1991). The S's were clearly divided into two groups, 3 S's whose x_E and x_F were about 175cm and 4 S's whose x_E and x_F were about 238cm. In both groups, x_E and x_F were surprisingly close to $x_C = 274.3$cm. All midpoints on the respective sides, Q_D, Q_E and Q_F, were either on the sides or slightly inside the triangle. In the second experiment, the S set Q_G and Q_H on the side between Q_A and Q_B. For the two groups of S's mentioned above, the results were the same. All Q's were inside the line connecting Q_A and Q_B (x_H, $x_G > 71.1$cm), and the gap $\Delta y = y_H - y_G > 0$ except one S. The average of Δy was about 6.4cm, which implies $K < 0$. The tendency of Q's being inside the triangle is consistent with this conclusion because the geodesics representing the perceptually straight sides are curved toward the center in EM. Of the exceptional S, $\Delta y = -0.25$cm is close to zero. This S, who was called "Euclidean observer", confessed later "I cannot forget the (Euclidean) theorem". We cannot estimate the value of $-K$ from the results of this experiment unless some mapping functions are assumed.

Watanabe (1996) conducted a similar experiment with an isosceles triangle in three different planes in the dark. Three fixed light points (Q_C, Q_A, Q_B) were presented in three different orientations; on the DP plane ($z = 0$), on a slanted plane and on the perceptually vertical plane. The results of horizontal case and vertical case are plotted in Figs.5.3,4 from Tables 3 and 5 in his article. Fig.5.3 shows the triangle and mean positions of bisecting points Q_D, Q_E, Q_F on the DP, where the coordinates (x, y, z) of Q_C, Q_A, and Q_B were respectively fixed at (133.5, 10, 0), (310, -120, 0), and (310, 113, 0) in cm. In contrast to the Blank's experiment, the vertex Q_C was closer to the S. Each of eight S's adjusted (x, y) of Q_D, Q_E, Q_F, and Q_G, Q_H three times. The overall means of eight S's are plotted in Fig.5.3. Positions of Q_D, Q_G, Q_H are shown above with a magnified scale. All Q's are inside of the triangle (Q_C, Q_A, Q_B) and the gap between Q_G, Q_H, $\Delta y = (y_H - y_G)$ is 0.4 cm. These results are in line with K < 0. However, this positive value of Δy may not be significant if tested. Using the Luneburg's mapping functions (2.3.1,2), Watanabe estimated the optimum values of the two constants, K = – 0.9, σ = 40.0.

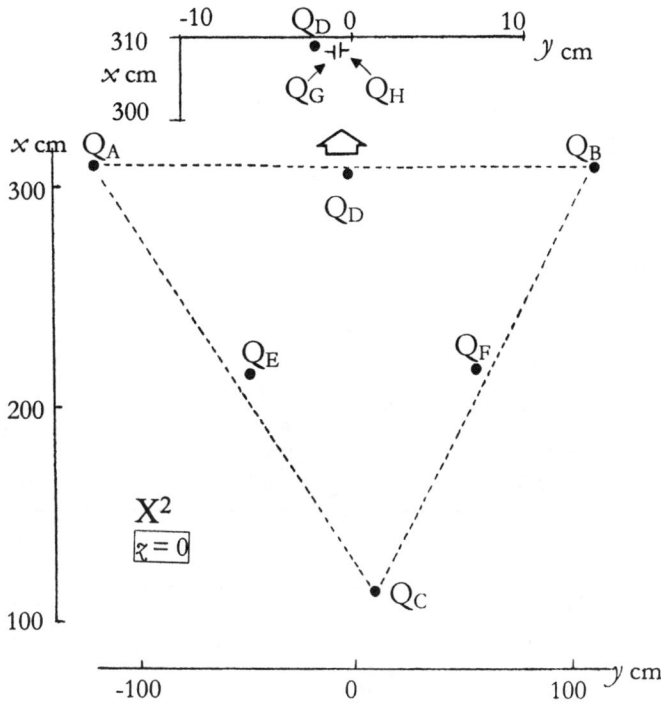

Fig.5.3 Watanabe's experiment (1996) on the horizontal DP

In the vertical case, Q_C was presented at (310 ± ε, 10, and 121) in cm while Q_A, Q_B were at the same positions on the DP (z = 0), (310, -120, 0) and (310, 113, 0). The first task of S was to adjust x of Q_C so that (Q_C, Q_A, Q_B) appeared on the HP plane perpendicular to the DP plane. The mean coordinates of Q_C were (290.6, 10, 121). Namely, the triangle (Q_C, Q_A, Q_B) was tilted in X^3 toward the S with $\tan^{-1}[(310 - 290.6)/121]$ = 10°. Then, the positions of Q_E, Q_F were adjusted in three directions x, y, and z whereas the positions of Q_D, Q_G, Q_H were adjusted in two directions, x and y. New eight S's participated in this experiment and the overall means of all S's were given in Table 5 of the original article. Fig.5.4 shows these results projected to the tilted plane (x, y, z') where z'

156 *Global Structure of Visual Space*

$= z /\cos 10°$. All Q's are either on the side or outside of the triangle (Q_C, Q_A, Q_B) and the overlap between Q_G, Q_H, $\Delta y = (y_H - y_G) = -0.7$ cm. Watanabe obtained K = 0.9 and σ = 6.0.

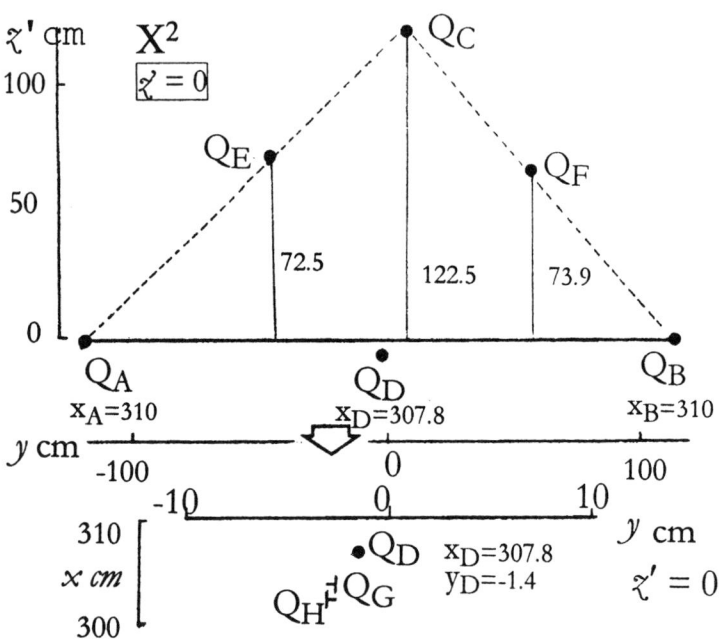

Fig.5.4 Watanabe's experiment on the vertical plane HP in VS

In the horizontal and vertical cases, $|K|$ is unusually close to 1.0. That K = -0.9 on the DP is consistent with the conclusion in Chapter 3. That K = 0.9 on the HP is not consistent with that K = 0 on HP (Chapters 3 and 4). In the slanted plane, the gap between Q_G, Q_H, $\Delta y = (y_H - y_G)$ is 0.2 cm and K = -0.45, σ = 36.0. Throughout the three cases, the discrepancies Δy between Q_G, Q_H are small and their relations to Q_D are ambiguous. Perhaps, these three perceived triangles (P_{VC}, P_{VA}, P_{VB}) are not too far from being Euclidean. As we already have seen in Secs. 3.1.2 and 3.2.2, when we try to fit equations having K and σ under these circumstances, the results become unstable.

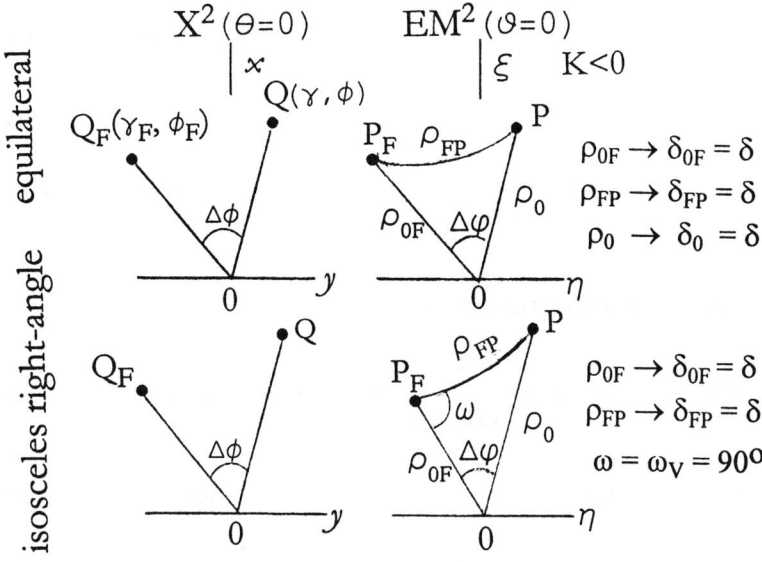

Fig.5.5 Higashiyama's experiment (1981)

Higashiyama (1981) performed a series of experiments with triangles in which the S was a vertex O. Though he used various kinds of triangles, two cases are discussed here; equilateral (regular) triangle and isosceles right-angle triangle (Fig.5.5). Two light points were presented on DP ($\theta = 0$) in a dark room. Point $Q_F(\gamma_F, \phi_F)$ was fixed and the S was asked to adjust the position of $Q(\gamma, \phi)$ so that (Q_F, O, Q) appeared either to be an equilateral triangle or to be an isosceles right-angle triangle. The configurations in the EM^2 when $K < 0$ are shown on the right side of Fig.5.5. When $K = 0$ or $K > 0$, the same notations hold. The only difference is the curve connecting P_F and P. It is a straight line when $K = 0$ and an arc convex upwards when $K > 0$. As will be explained in Sec.5.1.3, the following relationships hold with the separation angle $\Delta\varphi$ (> 0).

Equilateral triangle

$$K < 0 \quad \cos\Delta\varphi = \frac{1}{1 + 1/\cosh 2q\delta} > \frac{1}{2}, \quad \Delta\varphi < 60° \quad (5.1.4)$$

$K = 0 \quad \cos\Delta\varphi = \dfrac{1}{2}, \qquad \Delta\varphi = 60° \qquad (5.1.5)$

$K < 0 \quad \cos\Delta\varphi = \dfrac{1}{1 + 1/\cos 2q\delta} < \dfrac{1}{2}, \qquad \Delta\varphi > 60° \qquad (5.1.6)$

Isosceles right-angle triangle

$K < 0 \quad \tan\Delta\varphi = \dfrac{1}{\cosh 2q\delta} < 1, \qquad \Delta\varphi < 45° \qquad (5.1.7)$

$K = 0 \quad \tan\Delta\varphi = 1, \qquad \Delta\varphi = 45° \qquad (5.1.8)$

$K > 0 \quad \tan\Delta\varphi = \dfrac{1}{\cos 2q\delta} > 1, \qquad \Delta\varphi > 45° \qquad (5.1.9)$

With $Q_F(\gamma = 4.6°, \phi_F)$, where ϕ_F were varied in 9 ways from $\pm 5°$ ~ $\pm 45°$, he measured angle $\Delta\phi$ in X^2. Means of $\Delta\phi$ were plotted against $|\phi_F|$ in Fig.2 of his article (10 S's). All values of $\Delta\phi$ are less than 60° in the equilateral triangle and less than 45° in the isosceles right-angle triangle. However, these do not remain at a constant level as functions of $|\phi_F|$. Slightly positive trends are noticeable in both cases. If we assume, in accordance with Luneburg's mapping functions,

(1) $\rho_0 = g(\gamma)$ for all values of ϕ
(2) $\varphi = \phi$ and hence $\Delta\varphi = \Delta\phi$

the findings mean that $K(\phi_F) < 0$ and its value slightly varies according to $|\phi_F|$ in all the cases. Out of 180 (9×2×10) individual values of $K(\phi_F)$, 169 K's are negative in the equilateral triangle and 138 K's are negative in the isosceles right-angle triangle. By doing an additional experiment on lateral angles, he found out that (2) should be replaced by $\Delta\varphi = \Delta\phi^{1.03}$. Under this relationship, he obtained 119 $K < 0$ in the equilateral triangle and 75 K< 0 in the isosceles right-angle triangle (Table 2 in his article).

Foley (1972) reported two experiments on an isosceles right-angle triangle in the horizontal DP with the S as a vertex. In an experiment with small light points in the dark, Q_F was at (126, 0) in cm and $\Delta\phi$ was 29.8° for mean settings of 24 S's (each with three repetitions). Even if $\Delta\phi = \Delta\phi^{1.03}$, $\Delta\phi$ is clearly smaller than 45°. He repeated the same experiment in an illuminated environment with perspective cues. As Q's, white rods were presented against a black background on a black, white, and gray checked tablecloth, 16 cm below eye-level. With Q_F (80, 0, 17) in cm, $\Delta\phi$ was 39.9° with 12S's. In this case, $\Delta\phi = \Delta\phi^{1.03}$ is almost 45°. He wrote "geometry may approach Euclidean geometry with the introduction of cues to distance".

Higashiyama (1984) experimentally determined the locus of constant δ_0 and the relation between ϕ and ϕ. The S's were asked to assess δ_0 in terms of subjective meter (sm) and assess ϕ in terms of subjective degree. Using these results, he obtained individual values of K with nine S's. For isosceles right-angle triangles in which $|\phi_F|$ was varied from 5° to 40° in a 5° step, geometric means of K were between −0.50 and −0.78 (Table 3 in his article).

Eqs. (5.1.4 – 5.1.9) hold for any triangle like (P_{VC}, P_{VF}, P_{VP}) and the assertions on $\Delta\phi$ are evident from the well-known theorem; the sum of internal angles of a triangle is less than or equal to or larger than 180° according to K < 0 or K = 0, or K > 0. However, it is not possible to directly test this theorem in VS. In Fig.5.5, ω is set equal to 90° because the S adjusts the perceived angle ω_V to be a right angle, and VS and EM are known to be conformal (EM2 in Sec.2.2.2). We cannot directly measure ω_V and, as to this angle, VS and X is not conformal. If Q_C is not at O as in Fig.5.3, we have the same circumstances with all angles. One advantage of using the S as a vertex is that it is straightforward how to measure $\Delta\phi$ and how to relate it to $\Delta\phi$. Another advantage is that, once we have a scale value of perceptual radial distance δ_0, it opens a way to define numerical value of K.

5.1.3. *General Comments and Derivations*

The experiment of Blank (Fig.5.2) presupposes only a monotonic relation between VS and X. Without using the Luneburg's mapping functions, we can determine the sign of K, but not its value. His results

160 *Global Structure of Visual Space*

showed that the perceived layout $\{P_{Vi}\}$ on the frameless DP ($\vartheta = 0$) is of hyperbolic structure. According the experiment of Watanabe, it is either of hyperbolic structure or of Euclidean structure. Other experiments referred to in this section use the Luneburg's mapping functions directly or indirectly. All investigators admit that we can define K once a configuration $\{P_{Vi}\}$ is fixed. On the DP, $\{P_{Vi}\}$ tends to be hyperbolic but the value of K seems to depend on $\{P_{Vi}\}$. Most investigators see some problems in the Luneburg's mapping functions as defined in Eqs. (2.3.1) and (2.3.2). The problems will be pursued in the next section.

Eqs.(5.1.1) to (5.1.9) in Sec.5.1.2 may need some explanations. All are based upon Eqs.(2.4.7). Using notations of $\Delta P_{VA} P_{VB} P_{VC}$ in Fig.5.2, we can write Eqs.(2.4.7) as follows.

$$K = 0 \quad c^2 = a^2 + b^2 - 2ab \cos \omega_V \qquad (5.1.10)$$

$$K < 0 \quad \cosh c = \cosh a \cosh b - \sinh a \sinh b \cos \omega_V \qquad (5.1.11)$$

$$K > 0 \quad \cos c = \cos a \cos b + \sin a \sin b \cos \omega_V \qquad (5.1.12)$$

The last two formulae are related by the formula

$$\cos iX = \cosh X, \quad \sin iX = i \sinh X \qquad (5.1.13)$$

Eq.(5.1.1) when K = 0 is immediate.

$$d^2 = (\frac{a}{2})^2 + (\frac{b}{2})^2 - 2(\frac{a}{2})(\frac{b}{2}) \cos \omega_V \quad \text{from } \Delta P_{vc} P_{ve} P_{vF}$$

Combining it with (5.1.10), immediately we have

$$d^2 = \frac{c^2}{4}, \quad \text{hence Eq.(5.1.1)}$$

To obtain Eq.(5.1.2), K < 0, let us consider $\cosh^2 d$. Then, using the following three formulae,

$$\cosh^2 X - \sinh^2 X = 1$$

$$\sinh 2X = 2\sinh X \cosh X$$
$$\cosh 2X = 1 + 2\sinh^2 X = 2\cosh^2 X - 1$$

we have

$$\cosh^2 d = A + B - 2C$$

$$A = \cosh^2 \frac{a}{2}\cosh^2 \frac{b}{2} = (1+\sinh^2 \frac{a}{2})(1+\sinh^2 \frac{b}{2})$$

$$B = \sinh^2 \frac{a}{2}\sinh^2 \frac{b}{2}\cos^2 \omega_V$$

$$C = \cosh\frac{a}{2}\sinh\frac{a}{2}\cosh\frac{b}{2}\sinh\frac{b}{2}\cos\omega_V = \frac{1}{4}\sinh a \sinh b \cos\omega_V$$

From Eq.(5.1.11),

$$C = \frac{1}{4}(\cosh a \cosh b - \cosh c)$$
$$= \frac{1}{4}\{(1+2\sinh^2 \frac{a}{2})(1+2\sinh^2 \frac{b}{2}) - \cosh c\}$$

Putting all together, we have

$$\cosh^2 d = \frac{1}{2} + \frac{1}{2}\cosh c + \sinh^2 \frac{a}{2}\sinh^2 \frac{b}{2}(\cos^2 \omega_V - 1)$$

hence

$$\cosh 2d = \cosh c - 2\sinh^2 \frac{a}{2}\sinh^2 \frac{b}{2}\sin^2 \omega_V$$

and we have Eq.(5.1.2).

Eq.(5.1.3), K > 0, is obtained in a similar way or by applying Eq.(5.1.13) to cosh2d, we have

$$\cos 2d = \cos c - 2\sin^2 \frac{a}{2} \sin^2 \frac{b}{2} \sin^2 \omega_V$$

Eqs.(5.1.4) – (5.1.6) can be obtained by the use of Eqs.(5.1.11) – (5.1.12) in which c, a, b, and ω_V are replaced by δ_{FP}, δ_{0F}, δ_{0P}, and $\Delta\varphi$.

Equilateral triangles, $\delta_{FP} = \delta_{0F} = \delta_{0P} = \delta$

K = 0

$$\delta_{FP}^2 = \delta_{0F}^2 + \delta_{0P}^2 - 2\delta_{FP}^2 \cos\Delta\varphi$$

$$2\delta^2 \cos\Delta\varphi = 2\delta^2 - \delta^2 = \delta^2$$

Hence, $\cos\Delta\varphi = \dfrac{1}{2}$. Thus we have Eq. (5.1.1).

K < 0

$$\cosh 2q\delta_{FP} = \cosh 2q\delta_{0F} \cosh 2q\delta_{0P} - \sinh 2q\delta_{0F} \sinh 2q\delta_{0P} \cos\Delta\varphi$$

$$\cosh 2q\delta = \cosh^2 2q\delta - \sinh^2 2q\delta \cos\Delta\varphi$$

$$\cos\Delta\varphi = \frac{\cosh 2q\delta(1 - \cosh 2q\delta)}{1 - \cosh^2 2q} = \frac{\cosh 2q\delta}{1 + \cosh 2q\delta}$$

and we have (5.1.4).

K > 0

$$\cos 2q\delta = \cos^2 2q\delta - \sin^2 2q\delta \cos\Delta\varphi$$

$$\cos\Delta\varphi = \frac{\cos 2q\delta(1 - \cos 2q\delta)}{1 - \cos^2 2q\delta} = \frac{\cos 2q\delta}{1 + \cos 2q\delta}$$

and we have Eq.(5.1.6)

Eqs.(5.1.7)–(5.1.9) are obtained as follows.
In isosceles right-triangles, $\delta_{FP} = \delta_{0F} = \delta$, $\omega_V = 90°$.

$$K = 0 \quad \tan \Delta\varphi = \frac{\delta_{0F}}{\delta_{FP}} = 1$$

Using

$$\tan X = \frac{\sin X}{\cos X}, \quad \tanh X = \frac{\sinh X}{\cosh X}$$

we have

$$K < 0 \quad \tan \Delta\varphi = \frac{\tanh 2q\delta_{0F}}{\sinh 2q\delta_{FP}} = \frac{\tanh 2q\delta}{\sinh 2q\delta} = \frac{1}{\cosh 2q\delta}$$

$$K > 0 \quad \tan \Delta\varphi = \frac{\tan 2q\delta_{0F}}{\sin 2q\delta_{FP}} = \frac{\tan 2q\delta}{\sin 2q\delta} = \frac{1}{\cos 2q\delta}$$

5.2. Mapping Functions

When discussing the Luneburg's mapping functions, most theoreticians refer to two experiments by Foley (Blank, 1978; Lucas, 1983; Heller, 1997). Hence, these experiments are explained first. Then, theoretical discussions about mapping functions will be explored. Finally, the meaning of mapping functions will be discussed.

5.2.1. *Experimental Data*

The position of a stimulus point Q is characterized by either (γ, ϕ, θ) or (e_0, ϕ, θ) and that of the perceived point P_v in VS by (δ_0, ϕ_V, ϑ_V) (Fig.2.3). The basic idea in the first equation of the Luneburg's mapping function (2.3.1) is that the perceptual radial distance δ_0 is determined, not by the physical distance e_0, but by the convergence angle γ. It is through the radial distance ρ_0 in EM, as a function of γ, and the same $\rho_0 = g(\gamma)$ is assumed to hold for all directions (ϕ, θ). On the DP($\theta = 0$), loci of constant γ are called the Vieth-Müller circles (dotted circles in Fig.1.3). Hence, all points Q's on a Vieth-Müller circle are assumed to appear at the same radial distance $\delta_0(\gamma)$. This assumption was tested by an experiment carefully conducted by Foley (1966).

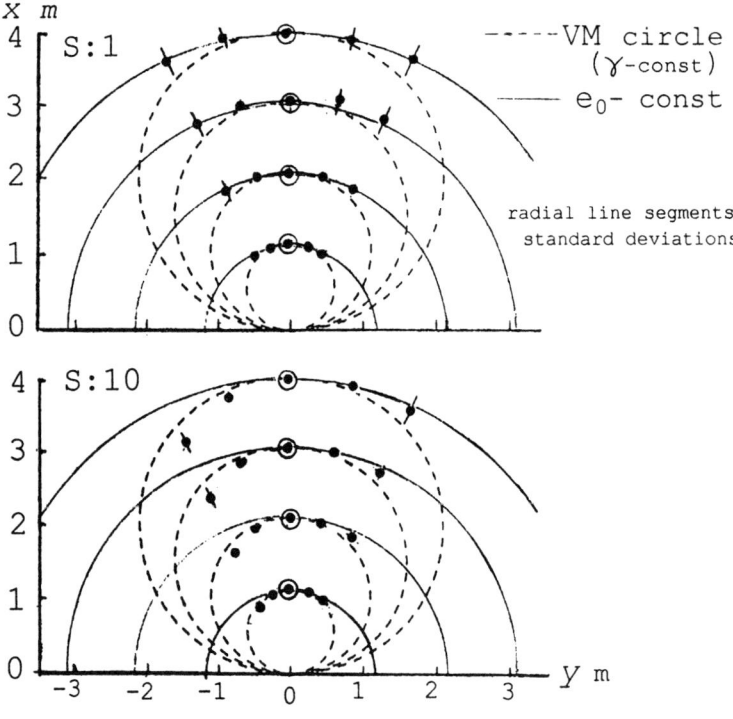

Fig.5.6 Foley's experiment (1966)

Five small light points $Q_i(\gamma_i, \phi_i)$ were presented in complete darkness on the DP, where $\phi_i = -24, -12, 0, 12,$ and 24 in degree. The central Q_0

was fixed at (x_0, 0, 0) and the experimenter moved Q_i along the direction ϕ_i until the S judged that its radial distance δ_{0i} appeared the same with δ_0 to the central Q_0. Once positions of the four $Q(x_i, \phi_i)$ were determined, the S saw five points lying on a half circle with the self at the center. The S was allowed to change any of the settings until this criterion was satisfied. The head of S was fixed and all Q's were adjusted in advance to appear with the same brightness. The position of Q_0 was varied in four ways, $x_0 = 1.2, 2.2, 3.2$, and 4.2 m. Ten college students served as S's. All had approximately normal acuity, uncorrected. With each $Q(x_0, 0, 0)$, each S repeated setting of each $Q(x_i, \phi_i)$ 35 times. Means of x_i of the last 25 settings were given in Table 1 of his article.

The results of two S's are replotted in Fig.5.6 from Fig.1 of Heller (1997). Vieth-Müller circles on which γ is constant are shown by dotted curves. Half circles are added on which the physical distance e_0 is constant. Radial line segments indicate standard deviation of settings. The mean settings of S1 are symmetric with regard to the x-axis whereas those of S10 are not. This type of asymmetry is often found in binocular vision experiments, which is ascribed to aniseikonia, *i.e.*, "anomally in the functional organization of the images from the two eyes" (Ogle 1964). In S1, points are almost on the half circles of constant e_0 in this range of ϕ. In S10, when the asymmetries are eliminated by taking means of x_i of $\pm \phi_i$, points are almost on the Vieth-Müller circles. There is a general tendency that points are outside the Vieth-Müller circles. Hence, as already discussed in Sec.5.2.1, the assumption that $\rho_0 = g(\gamma)$ is the same for all directions (ϕ) on the DP does not hold strictly. The Luneburg's mapping function (2.3.1) specifies the form of $g(\gamma)$. To test the assumed form $g(\gamma)$ for Q's on or close to the x-axis, we need to have a scale expressing the latent variable $\delta_0(x, 0, 0)$. Foley (1978, 1980) proposed on the basis of experiments a formula that is different from the one stated above. His formula was experimentally disproved by Lukas (1983).

The second mapping function of Luneburg, Eq.(2.3.2), assumes that two directional angles (φ, ϑ) in EM are equal to the corresponding angles (ϕ, θ) in X. Under this condition, perception of direction is veridical because two directional angles (φ_V, ϑ_V) in VS are equal to (ϕ, θ) in X (Eq.2.3.6). In the DP($\theta = 0$), loci of constant ϕ are called the hyperbolae of Hillebrand for the reason to be stated in Sec.5.2.2. It is

166 *Global Structure of Visual Space*

generally accepted that, in a frameless VS, all Q's on a hyperbola(ϕ) are perceived on a constant direction (ϕ_V). It does not necessarily imply that $\phi_V = \phi$ in the sense that when Q_A in the direction ϕ_A (> 0) and Q_B is in the direction $2\phi_B$ (> 0), then P_B appears in VS in the direction of twice right as that of P_A, *i.e.*, (ϕ_V of P_B) = 2 (ϕ_V of P_A). This is the point to which a doubt was cast in Sec.5.1.1. Foley reported an experiment at the 73rd Annual Convention of American Psychological Association (1972). He scaled perceived size of physical angle $\Delta\phi$ formed by small light points in a dark field by methods of fractionation, magnitude estimation, and magnitude production, and found the scaled size $\Delta\phi_V = 1.1\Delta\phi$.

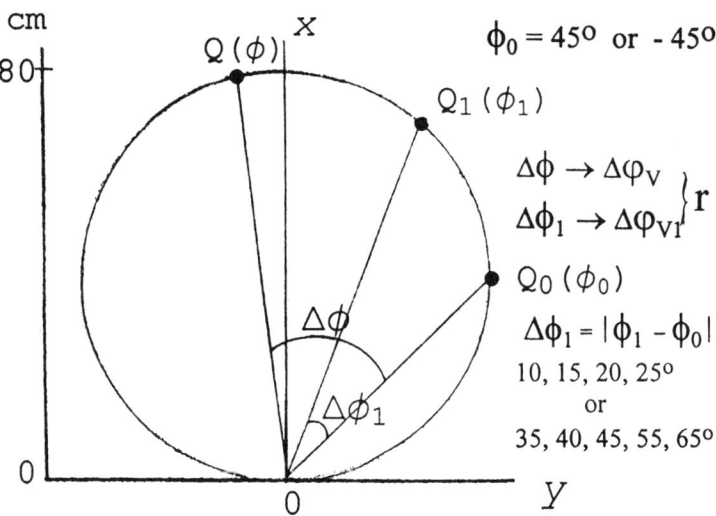

Fig.5.7 Higashiyama's experiment (1981)

A similar result was obtained by Higashiyama (1981). Three small light points Q's were presented in total darkness on a circle of diameter 80cm that touched the origin 0 in the DP-plane (Fig.5.7). A standard pair, $Q_0(\phi_0)$ and $Q_1(\phi_1)$, were fixed to form a standard physical angle $\Delta\phi_1$ and the S adjusted ϕ of Q so that the perceptual angle $\Delta\phi_V$ from Q_0 to Q appears r times the perceived standard angle $\Delta\phi_{V1}$ from Q_0 to Q_1; $\Delta\phi_V = r \Delta\phi_{V1}$.

doubling r = 2,   tripling r = 3
bisection r = 1/2   trisection r = 1/3

The head of S was fixed. The position of Q_0 was varied in two ways as shown in Fig.5.7. The standard angle $\Delta\phi_1$ was varied four ways in doubling and tripling (10, 15, 20, 25°) and five ways in bisection and trisection (35, 40, 45, 55, 65°). Eight S's participated in this experiment. He assumed a power function between $\Delta\phi_V$ and $\Delta\phi$ and estimated the value of the exponent β in each $\Delta\phi_1$ from each Q_0 by the method given by Stevens (1957).

$$\Delta\phi_V = k\Delta\phi^\beta, \quad \Delta\phi_{V1} = k\Delta\phi_1^\beta, \quad k > 0$$

$$r = \frac{\Delta\phi_V}{\Delta\phi_{V1}} = \left(\frac{\Delta\phi}{\Delta\phi_1}\right)^\beta, \quad \log r = \beta \log \frac{\Delta\phi}{\Delta\phi_1}$$

Hence,

$$\beta = \frac{\log r}{\log \frac{\Delta\phi}{\Delta\phi_1}} \tag{5.2.1}$$

Means of β over S's and $\Delta\phi_1$ are given in Table 3 of his article. They are fairly stable across the four conditions, from 0.99 (bisection) to 1.09 (tripling) with the mean = 1.03. In this way, he obtained $\Delta\phi_V = \Delta\phi^{1.03}$ that was referred to in Sec.5.1.2. According to him, this equation and the Foley's equation $\Delta\phi_V = 1.1\Delta\phi$ yield indistinguishable values in the range of angles under discussion.

5.2.2. Theoretical Considerations

Though we cannot expect the mapping functions to hold exactly in the form as assumed by Luneburg, it will be of interest to discuss implications of Eqs. (2.3.1, 2). Suppose a situation that a stimulus point

168 *Global Structure of Visual Space*

Q(γ, φ, θ) is presented in the physical space X^3 and the perceived point in a frameless VS^3 and its representation in Euclidean map EM^3 are respectively denoted as $P_V(\delta_0, \varphi_V, \vartheta_V)$ and $P(\rho_0, \varphi, \vartheta)$. Consider in each space a line element from the point and a small plane Ω around the point that is orthogonal to the line of sight (Fig.5.8I). The line element can be in any direction and it is decomposed into two components, one along the line of sight and the other on Ω; ds into $d\delta_0$ and $d\varepsilon_V$ in VS, $d\rho$ into $d\rho_0$ and $d\varepsilon_{EM}$ in EM, and de into de_0 and $d\varepsilon_X$ in X.

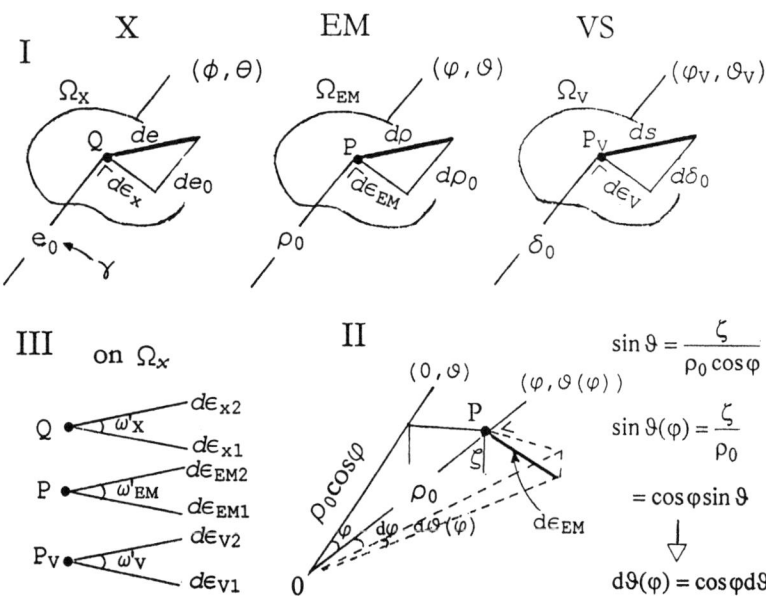

Fig.5.8 Line elements in X, EM, and VS

VS $$ds^2 = d\delta_0^2 + M(\delta_0)^2 d\varepsilon_V^2$$
$$d\varepsilon_V^2 = (d\varphi_V^2 + \cos^2\varphi_V d\vartheta_V^2) \tag{5.2.2}$$

EM $$d\rho^2 = d\rho_0^2 + \rho_0^2 d\varepsilon_{EM}^2$$
$$d\varepsilon_{EM}^2 = (d\varphi^2 + \cos^2\varphi d\vartheta^2) \tag{5.2.3}$$

X $\quad de^2 = de_0^2 + e_0^2 \, d\varepsilon_x^2$
$\quad\quad d\varepsilon_x^2 = (d\phi^2 + \cos^2\phi \, d\theta^2)$ (5.2.4)

Fig.5.8II illustrates, with regard to Eq.(5.2.3), how $d\varepsilon$ in each equation is defined, where $d\vartheta(\varphi) = \cos\varphi \, d\vartheta$ is obtained by expanding the sin functions into the power series from $\sin\vartheta(\varphi) = \cos\varphi \sin\vartheta$. Since EM and X are Euclidean, $d\varepsilon$'s increase in size according to distance from the origin for fixed $d\varphi$ and $d\vartheta$, etc. It is not clear in VS how $d\varepsilon_V$ is related δ_0 and the scale factor $M(\delta_0)$ is used in Eq.(5.2.2). From Eq.(2.2.3),

$$\frac{d\delta_0}{d\rho_0} = \frac{1}{G(\rho_0)^2} \quad (5.2.5)$$

where

$$K < 0 \quad G(\rho_0)^2 = \left(1 - \frac{(-K)^2}{4}\right)\rho_0^2$$

$$K = 0 \quad G(\rho_0)^2 = \rho_0^2$$

$$K > 0 \quad G(\rho_0)^2 = \left(1 + \frac{K^2}{4}\right)\rho_0^2$$

It was explained in Sec.2.4.2 that two directional angles from the self are the same in EM and VS, *i.e.*, $\varphi = \varphi_V$, and $\vartheta = \vartheta_V$. Hence, $d\varepsilon_V = d\varepsilon_{EM}$. In accordance with Eq.(2.2.5), if we define the scale factor as

$$M(\delta_0) = \frac{\rho_0}{G(\rho_0)^2} \quad (5.2.6)$$

then

$$ds = G(\rho_0)^{-2} d\rho \quad (5.2.7)$$

Eq.(5.2.6) is equivalent to putting $M(\delta_0) = \delta_0$. That ds and $d\rho$ are proportional implies that, between VS and EM, non-directional angles are preserved also. Suppose two line elements from P_V in any direction in VS^3, ds_1 and ds_2 with an angle ω_V, and denote their representations in EM^3 as $d\rho_1$ and $d\rho_2$ from P with the angle ω_{EM}.

$$\cos \omega_V = \frac{\langle ds_1\ ds_2 \rangle}{\sqrt{\langle ds_1\ ds_1 \rangle}\sqrt{\langle ds_2\ ds_2 \rangle}}$$

$$\cos \omega_{EM} = \frac{\langle d\rho_1\ d\rho_2 \rangle}{\sqrt{\langle d\rho_1\ d\rho_1 \rangle}\sqrt{\langle d\rho_2\ d\rho_2 \rangle}}$$

where $\langle \rangle$ means the scalar product. Inserting (5.2.7) in the first equation, we have the second equation and hence $\omega_V = \omega_{EM}$, which means that any angle in VS^3 is represented in EM^3 without any distortion. In general, angles are preserved between two spaces when line elements are proportional.

Suppose the corresponding two line elements in X^3, de_1 and de_2 with the angle ω_X in X^3. Radial distances ρ_0 in EM^3 and e_0 in X^3 are not directly related. These are related only through the convergence angle γ. Hence de is not reduced to a form that is proportional to $d\rho$ and to ds. Namely the angle ω_X is not equal to ω_V in general. However, consider two line elements on the respective planes Ω; $d\varepsilon_{X1}$, $d\varepsilon_{X2}$ with an angle ω_X' on Ω_X, $d\varepsilon_{EM1}$, $d\varepsilon_{EM2}$ with the angle ω'_{EM} on Ω_{EM}, and $d\varepsilon_V$, $d\varepsilon_{V2}$ with an angle ω'_V on Ω_V (Fig.5.8III). For these, $de_0 = d\rho_0 = d\delta_0 = 0$, and hence,

$$ds = M(\delta_0)\, d\varepsilon_V, \quad d\rho_0 = \rho_0\, d\varepsilon_{EM}, \quad de = e_0\, d\varepsilon_X$$

Under the mapping functions (2.3.1), $\varphi_V = \varphi = \phi$, $\vartheta_V = \vartheta = \theta$ and hence $d\varepsilon_V = d\varepsilon_{EM} = d\varepsilon_X$. Namely all line elements are proportional and $\omega'_V = \omega'_{EM} = \omega'_X$. The angle ω'_X is the retinal image of a physical angle ω_X of a stimulus configuration in X^3 or an angle ω_X on a frontoparallel plane. Hence the mapping functions (2.3.1) imply that, insofar as this angle is concerned, we perceive it as projected on the retina. Up to this point, the first mapping function Eq.(2.3.1) has not been used.

If we use Eq.(5.2.6) and Eq.(2.3.8) based on Eq.(2.3.1)

$$\frac{d\delta_0}{d\gamma} = \frac{-\sigma\rho_0}{G^2}$$

and we can write Eq.(5.2.2) in the form (Eschenburg, 1980).

$$ds = \frac{\rho_0}{G^2}\left(\sigma^2 d\gamma^2 + d\varepsilon_X^2\right)^{-2} \tag{5.2.8}$$

This remains the same for any $Q(\gamma, \phi, \theta)$ if γ is constant as on the Vieth-Müller circle. When Luneburg introduced the mapping functions (Eqs.2.3.1, 2), he may have had in his mind their implications as stated above. After 50 years, Heller (1997) extended deep theoretical discussions on mapping functions for Q on the horizontal DP in X.

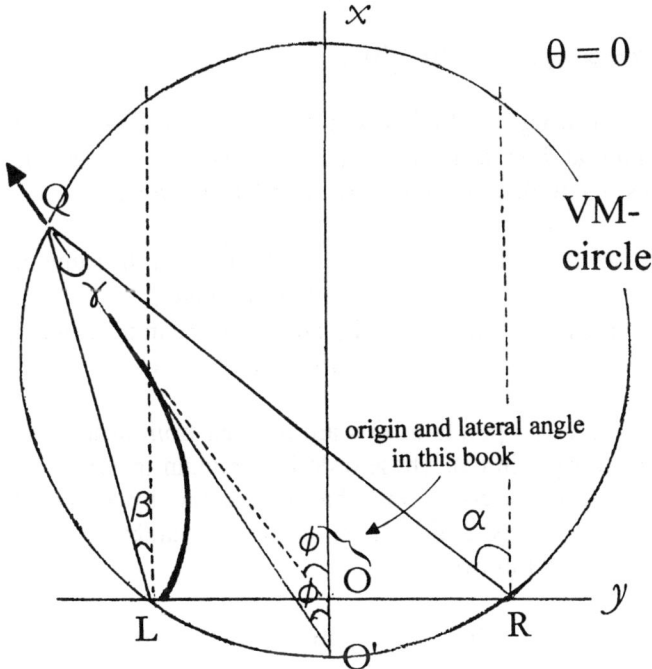

Fig.5.9 Hillebrand hyperbola and coordinates used by Heller (1997)

172 *Global Structure of Visual Space*

Luneburg defined the direction ϕ of Q from O' on the Vieth-Müller circle, which was followed by Heller (Fig.5.9). Then, the locus on which this ϕ is constant is the Hillebrand hyperbola (the thick curve) that passes through the eye. On the other hand, in this book, ϕ is defined from the origin O in the center of two eyes (Figs.1.3, 2.1 etc.). According to this definition, the locus on which ϕ is constant is simply a straight line as shown by the dotted radial line from O'. This is close to the asymptote of the Hillebrand hyperbola, the continuous line form O. In most parts of this book, we are not concerned about the difference between the two loci and the difference between O and O'. These differences are negligible in practice and Q on the curved part of the Hillebrand hyperbola is not dealt with in experiments. Eqs.(2.3.3a, b) in Sec.2.3.1 refers to ϕ defined from O'.

Heller defines the position of Q on the horizontal DP by two parameters (α, β) (Fig.5.9). These (α, β) are different from (α, β) used by Luneburg. From Heller's (α, β),

$$\gamma = \alpha - \beta, \quad \phi = (\alpha + \beta)/2 \tag{5.2.9}$$

and the following set of Q's is defined, $S_Q = \{(\alpha, \beta) \in \mathbf{I}^2\}$ where $\mathbf{I}$ is the open interval between $-\pi/2$ and $\pi/2$. He introduces on S_Q binary relations $\sim\prec$ to denote the following order judgments

(1) $(\alpha, \beta) \sim\prec_f (\alpha', \beta') \Leftrightarrow$ [Q appears to be at the same distance with or far than Q']
(2) $(\alpha, \beta) \sim\prec_r (\alpha', \beta') \Leftrightarrow$ [Q appears to be in the same direction with or right of Q']

That is, perceived points of Q's are ordered from near to far in (1) and from left to right in (2). These orderings remain invariant when fixation changes. Under a constraint, that is testable by order judgments if necessary, he showed that there exist real-valued, strictly increasing functions f, g as follows.

(1) $(\alpha, \beta) \sim\prec_f (\alpha', \beta')$ iff $f(\alpha) - g(\beta) \geq f(\alpha') - g(\beta')$

(2) $(\alpha, \beta) \sim\prec_r (\alpha', \beta')$ iff $f(\alpha) + g(\beta) \geq f(\alpha') + g(\beta')$ (5.2.10)

where iff means "if and only if". To cope with experimental results described in Sec.5.2.1, he defined

$$\Gamma = f(\alpha) - g(\beta), \qquad \Phi = [f(\alpha) + g(\beta)]/2 \qquad (5.2.11)$$

and postulated that δ_0 and φ_V of the perceived point of Q are determined by Γ and Φ. This is a generalization of the Luneburg's mapping functions in which δ_0 and φ_V are determined by γ and ϕ.

Solving functional equations, Aczél et al. (1999) reached the following conclusions. An arbitrary constant is denoted by τ.

(A) It is called γ-shift invariance that $(\alpha, \beta) \sim \prec_{f,r}(\alpha', \beta')$ iff $(\alpha+\tau, \beta-\tau) \sim \prec_{f,r}(\alpha'+\tau, \beta'-\tau)$. The functions f and g that satisfy this condition for all $(\alpha, \beta) \in S_Q$, $(\alpha+\tau, \beta-\tau) \in S_Q$ are linear forms or exponential forms only.

(B) It is called ϕ-shift invariance that $(\alpha, \beta) \sim \prec_{f,r}(\alpha', \beta')$ iff $(\alpha+\tau, \beta+\tau) \sim \prec_{f,r}(\alpha'+\tau, \beta'+\tau)$. The functions f and g that satisfy this condition for all $(\alpha, \beta) \in S_Q$, $(\alpha+\tau, \beta+\tau) \in S_Q$ are also linear forms or exponential forms only.

(C) The functions f and g that satisfy both (A) and (B) are linear forms only.
$$f(\alpha) = a\alpha + b, \quad g(\beta) = c\beta + d \qquad (5.2.12)$$

where $a > 0$, $c > 0$, b and d are constants. Because of these constants, Eq.(5.2.11) is more flexible than Eq.(5.2.9). In Fig.4 of their article are shown loci of constant δ_0 and loci of constant φ_V from Eq.(5.2.12), in particular $f(\alpha) = \alpha$ and $g(\beta) = 1.2\beta$ (aniseikonia). Their shapes are similar to the asymmetric results of S10 in Fig. 5.6.

5.2.3. Roles of Mapping Functions

According to Luneburg (1947, p.21), the following transformations were used by Ames in constructing his famous distorted rooms. If a given rectangular room and a distorted room are binocularly observed from the respective vantage points O and O', the two appear almost identical in size, shape, and position relative to the observer. A points $Q(\gamma, \phi, \theta)$ on the wall of the rectangular room and the corresponding point $Q'(\gamma', \phi', \theta')$ on the wall of the distorted room are related by

(1) $\gamma' = \gamma +$ (a constant)
(2) $\phi' = \phi$, $\theta' = \theta$

This is a special case of the class of transformations called *iseikonic*. The curved wall of the distorted room and the plane wall of the rectangular room have the same "binocular characteristics"; $d\gamma' = d\gamma$, $d\phi' = d\phi$, $d\theta' = d\theta$. The only difference is the value of γ. Luneburg wrote "the assumption that the convergence of the lines of sight is immaterial for binocular space perception then leads to the consequence that the observer sees the equivalent curved wall as plane if a suitable pattern on the wall induces him to this interpretation". The organization of a configuration $\{Q_i\}$ as a whole has even more priority over the effect of γ_i of individual points Q_i in generating the perceived pattern $\{P_{Vi}\}$. The primary factor to determine where individual Q_i appear as P_{Vi} in VS is, not their $(\gamma_i, \phi_i, \theta_i)$, but their relative positions in the total pattern in $\{P_{Vi}\}$.

When a perceived pattern $\{P_{Vi}\}$ with geometrical structure, such as P- and D-alleys and H-curves, is generated from $\{Q_i\}$ in a frameless X, it is true that the theoretical equations projected from EM^2 to X^2 through the Luneburg's mapping functions (Eqs.2.3.1,2) give excellent results (*e.g.*, Fig.2.1). However, it does not exclude the possibility that there are other sets of mapping functions that will yield the same or even better fits. In this sense, from the viewpoint of examining the geometrical structure of a perceived pattern, it is not too important to determine the precise form of point-to-point mapping functions that are expected to hold for any pattern $\{Q_i\}$. Making explicit the exact form of mapping functions for a simple ego-centrically localized $\{Q_i\}$ may be of interest to understand the physiological process that starts from a local stimulation in the retina and ends at the brain excitation underlying the perception of a points in a frameless VS (which is called primitivier Sehraum by Heller). The physiological aspect is not of primary concern of this book, however.

In Sec.3.2, an approach was explained that directly constructs the configuration $\{P_j\}$ in EM from a given $\{Q_i\}$ in X without intervening any *a priori* assumed mapping functions. Once we have a $\{P_j\}$ in this way, we can have the *a posteriori* mapping functions for this pattern by comparing $\{P_j\}$ and $\{Q_i\}$. To have mapping functions for $\{Q_i\}$ in advance is necessary when we need to predict what visual pattern $\{P_{vi}\}$ will be generated by $\{Q_i\}$. Designing a distorted room is such a case and similar situations will be encountered in virtual reality. If quantitative prediction is not a matter of concern, however, it is not a problem to know mapping functions, not in advance, but with $\{P_j\}$ for a geometrically structured $\{P_{vi}\}$. For non ego-centric spatial layouts as discussed in Chapter 4, the Luneburg's mapping functions are out of use.

5.3. Experimental Tests of Properties of VS as an R

In Sec.5.2, experiments and theoretical discussion on the Luneburg's mapping functions are reviewed. In this section, experiments are reviewed in which piecemeal tests are tried on properties that are implied in the postulate that VS is regarded as an R, Riemannian space of constant curvature. To define a geodesic in Riemannian space, it is necessary to generate it from infinitesimal element ds (Sec.2.4.1). If we start from the fact that we see straight lines in VS, however, their counterparts in Riemannian space need not be defined in this way. It was directly assumed that a latent line segment δ in VS can be regarded as a geodesic in Riemannian space and that δ is captured as a manifest variable d (Sec.2.2.1). Riemannian space is locally Euclidean and to assume that VS is locally Euclidean is not inconsistent with our experience. Furthermore, it was assumed that a configuration $\{P_{vi}\}$ is perceived in VS, and the curvature K remains the same within the configuration. All experimental results were interpreted on this assumption. However, to postulate VS as an R has various implications that have not yet been discussed. Some of these are experimentally testable.

5.3.1. *Two Experiments of Foley*

One of such implications is that VS is *Desargusian* (Blank, 1958). This property may be stated as follows with regard to a triangle (A,B,C) in

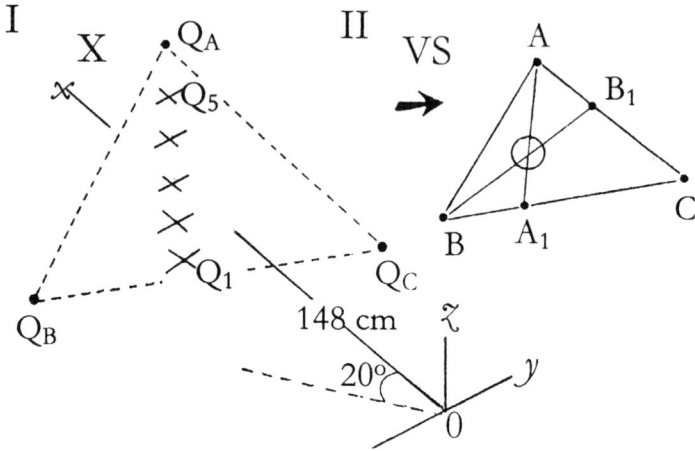

Fig.5.10 Foley's experiment on Desargusian property (1964a,b)

Fig.5.10II. If the perceived triangle on this plane is of constant K, given two line segments from one vertex to an arbitrary point on the opposite side (as A to a point A1 on BC and B to a point B1 on AB in Fig.5.10II), the two line segments intersect. Foley (1964a,b) tested this property with two triangles. Fig.5.10I is a schematic illustration to show the principle of the experiment. A configuration of light points $\{Q_i\}$ was presented in the dark. Each stimulus point Q_i consisted of a pair of polarized lights on a horizontal bar in the frame at $x = 148$ cm which were observed through two polarizers in front of the eyes. The intensity of light points was held constant at a level at which "these were clearly seen, but were not so bright as to illuminate the frame or interference with the stereoscopic effect". Of each Q_i, its z_i is fixed by the height of the horizontal bar, and its y_i and x_i were adjustable by changing its position on the bar and by changing the lateral separation of the stereoscopic lights. Each of the two triangles, $\{Q_i\}$, $i =$ A, B, C, was on a plane not perpendicular to the line of sight (Fig.5.10I).

Into the $\{Q_i\}$, lights to define line segments were introduced, one by one. First, (x, y) of Q_1 were adjusted so that A1 appeared on the side BC.

Then, the other four Q's were adjusted to define the segment connecting A and A1 (Fig.5.10II). The S was instructed to move the eyes, with the head fixed, over the entire configuration to use a bracketing procedure in setting the positions of Q's. The S was allowed to change the position of any light at any time during the construction of the segment. In the physical space X, Q_1 is not necessarily on the line connecting Q_B and Q_C, and the series Q_1-Q_5 is not necessarily straight.

In order to define the line segment B-B1 in VS, Q_5 was adjusted so that B1 appeared on the side AC. Then, other Q's were adjusted in a similar way. All necessary control was taken as to the practice, the order of settings, and the stability of settings in each session. The experiment was continued until 24 settings were obtained with each Q_i from each S. The results of three S's in each triangle are given in the original articles, one in terms of cm and the other in terms of visual angle. The minimum physical distance in the z-direction between two series of Q's, one yielding A-A1 and the other yielding B-B1, at the position indicated by a circle in Fig.5.10II was in the range from 0.014 to 0.162 cm. Notice that when one series was being adjusted, the other series was not visible. According to a statistical test, the distance was significantly larger than zero in two out of six cases (two triangles and three S's). However, it was demonstrated by an additional experiment that even the S who had yielded the largest difference was not able to discriminate whether the two segments were intersecting or not intersecting.

If a figure like Fig.5.10II on a large cardboard is presented in an illuminated environment, no matter how it is oriented toward the S, all sides of (ABC) will appear as straight lines and two straight lines, A-A1 and B-B1, will appear to intersect on the cardboard. In other words, VS in the neighborhood of the self is Desargusian. The Foley's experiment shows that the same is true with the configuration of small light points $\{P_{Vi}\}$ perceived on a plane in a frameless VS[3] even when A-A1 and B-B1 are independently determined.

In another article, Foley (1972) reported an experiment on another implication of the postulate that VS is an R. Of two stimulus configurations in Fig.5.11, either the smaller one $\{Q_A, Q_B, Q_C\}$ (circles) or the larger one $\{Q'_A, Q'_B, Q'_C\}$ (diamonds) was presented at a time. All Q's were on the horizontal DP and their intensities were the same level as those in the experiment in Fig.5.10. Q_A (Q'_A) was fixed (filled symbols). The S adjusted positions of Q_B (Q'_B) and Q_C (Q'_C) so that the

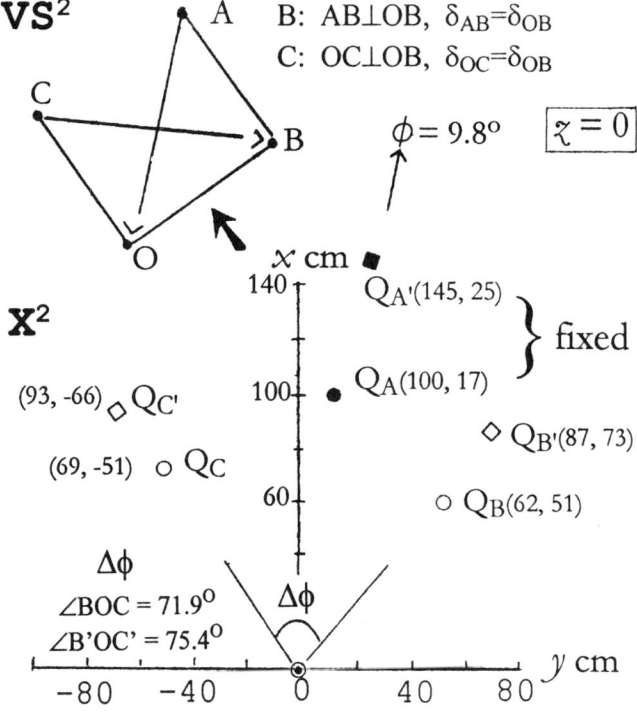

Fig.5.11 Foley's experiment on congruence (1972)

perceived pattern {A(A'), B(B'), O, C(C')}, including the S as O, satisfied conditions given in the inset in Fig. 5.11; (ABOC) consists of two isosceles right-angle triangles (ABO) and (COB) having the side OB in common. When the smaller configuration was adjusted, after setting Q_B and Q_C, the S was allowed to readjust either or both settings, and then was asked to judge which of OA and BC appeared the longer and to judge that ratio of their length. The procedure was the same when Q'_B and Q'_C of the larger configuration were adjusted.

Fig. 5.11 shows positions of Q_B, Q_C, and Q'_B, Q'_C, (median settings of 24 S's). Both configurations, smaller and larger, gave about the same $\Delta\phi$ (the angle in X^2 corresponding to $\angle BOC$), which implies that K is the same for the smaller and larger patterns (ABOC). That $\Delta\phi$ is much smaller than 90° implies that $K < 0$ (Sec.4.3.3). However, the results of distance judgments are contradictory to the hypothesis that the VS is of constant K. Irrespective of the sign of K, if K is constant, (ABO) and (COB) must be congruent and $\delta_{0A} = \delta_{BC}$. However,

(ABOC)	smaller	larger
percentage of $\delta_{BC} > \delta_{0A}$	87	79
ratio $\delta_{BC} / \delta_{0A}$	1.20	1.23

The smaller and larger patterns exhibit the same results with regard to $\Delta\phi$ and distance judgments. In both cases, setting Q_B (Q'_B) and Q_C (Q'_C) according to the criteria given in the inset does not lead to two congruent isosceles right-angle triangles.

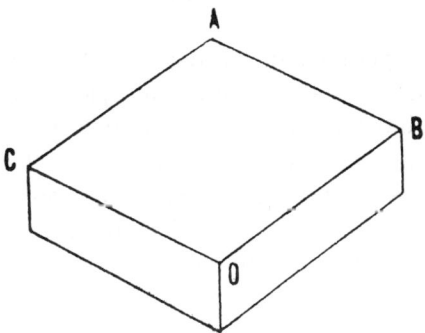

Fig.5.12 A square viewed from an oblique direction

In Fig.5.11, (ABOC) in the inset VS^2 shows the pattern viewed from above. The pattern that the S sees from position O in the process of setting is not this form; A, B, C are on the horizontal DP. It is not a straightforward task to place Q_B and Q_C so as to satisfy the specified criteria in the inset when these are on the DP. Foley mentioned, "the S was reminded not to substitute other criteria which might seem to be equivalent. A preliminary investigation had indicated that, in the absence of such an instruction, the tendency to make the configuration symmetrical, like a square, sometimes overrode the actual instructions".

Several comments may be made about this experiment, but two will suffice. One is the difficulty or ambiguity inherent in constructing a right angle when viewed from an oblique direction. Suppose that a figure, as shown in Fig.5.12, is drawn on a sheet of paper and presented at an angle to the line of sight, 90° or closer to zero. When asked about $\angle ABO$ or $\angle BOC$, people are apt to say it to be 90°, even when it is stressed that the question refers to the angle on the sheet of paper. The other is concerned with the idea that attending to equality and perpendicularity of two sides leads to two isosceles right-angle triangles. This is a bottom-up approach, so to speak. It would be interesting to try a top-down approach. For instance, Q_A and Q_B are fixed to define a triangle (ABO). It does not need be an isosceles right-angle triangle. Then the S is asked to locate Q_C so that (COB) appears to be congruent with (ABO) and then asked to judge whether $\delta_{BC} = \delta_{0A}$ or $\angle BAO = \angle BCO$.

5.3.2. Bottom-up Experimental Approaches

Zimmer (1998) experimentally tested one of the most basic conditions for regarding that VS is a geometrical space. Let a, b, c, and d be arbitrary four points that are on a one-dimensional continuum C in the order as shown in Fig.5.13I. Let us denote by $a|b|c$ that b is between a and c. Betweenness is an order relation that precedes metric relation. Suppes enumerated five axioms that ternary relations on a one-dimensional betweenness structure must satisfy (Suppes *et al.*, 1989). In addition to some trivial axioms such as "If $a|b|c$, then $c|b|a$", he gave

Axiom 3   If $a|b|c$ and $b|d|c$ then $a|b|d$
Axiom 4   If $a|b|c$ and $b|c|d$, and $b \neq c$ then $a|b|d$

Zimmer tested these two conditions with light points Q's appearing collinear in the dark. Each Q was a small red light, clearly visible with no halo. All Q's were on the DP($\theta = 0$). Herein, only one experiment to test Axiom 3 is explained.

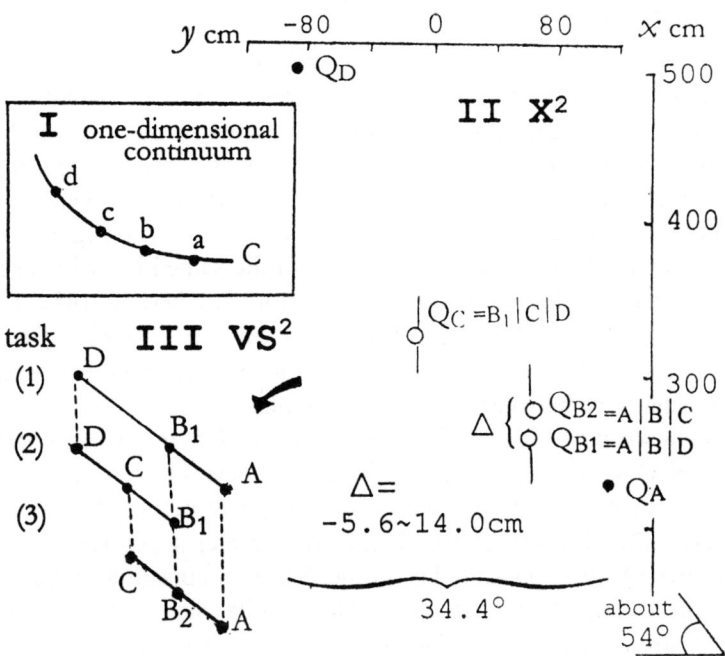

Fig.5.13 Schematic illustrations of betweenness settings

In Fig.5.13II, Q_A and Q_D were fixed whereas Q_B and Q_C were movable in the direction of x-axis. The S performed three tasks. (1) (Q_A, Q_B, Q_D) were presented and Q_B was set to such a position Q_{B1} that A, B_1, and D appear collinear (A|B_1|D). (2) (Q_{B1}, Q_C, Q_D) were presented and Q_C was set so that B_1, C, and D appear collinear (B_1|C|D). (3) (Q_A, Q_{B2}, Q_C) were presented and Q_{B2} was adjusted so that A, B_2, and C appear collinear (A|B_2|C). In each setting, the S was asked to imagine a thread connecting two outer Q's and to decide whether Q in the middle was on it or not. The position of Q was adjusted by the experimenter step by step in accordance with the judgment of S (the adaptive procedure in Falmagne, 1985) until it was judged to appear on the thread. Each of nine S's repeated each setting eight times. The experiment was carried out with extreme care. For instance, before the experimentation, the S was completely disoriented with regard to the physical structure of the room.

Three points, e.g., (A, B_1, D), are adjusted to be collinear in VS^2, which does not mean that (Q_A, Q_{B1}, Q_D) must be collinear in X^2. However, if Axiom 3 holds, Q_{B2} and Q_{B1} must coincide. There are large individual differences as to the mean difference of x, $\Delta = (Q_{B2} - Q_{B1})$ in cm (Fig.5.13II). In five out of nine cases, Δ is significantly different from zero (significant level $\alpha = 0.1$). The betweenness axiom does not seems to hold with this series of Q' that spans a visual angle of 34.4°. The same conclusion was reached by an experiment on Axiom 4. It is not because the series of Q's was tilted about 54° with regard to the x-axis. Zimmer performed an experiment of the same kind with a series of Q's that is parallel to the y-axis to test Axiom 4, where Q_A (412, 88, 0) in cm ($\phi = 12.0°$) and Q_D(412, –140, 0) in cm ($-\phi = 18.8°$). Because Δ ranges from –24.4 to 5.9 cm and is significant in four out of six cases ($\alpha = 0.1$), again the betweenness axiom is violated.

As shown by P-alley and H-curves in Fig.2.1, the S can construct a series of Q's so as to appear collinear. If a series (Q_A, Q_B, Q_C, Q_D) are simultaneously presented and adjusted to appear collinear, the S will see that (A|B|D), (B|C|D), and (A|B|C). In Zimmer's experiment, a set of only three Q's was presented at a time. Each set constitutes a geometrical pattern, collinearity in her case. Her results show, however, that the three patterns are not consistent with each other. This is a difficulty inherent in bottom-up approaches.

A series of experiments concerning the basic structure of VS have been conducted by Koenderink and his associates. One of these experiments, exocentric pointing in outdoor field with full cues, was discussed in Sec.4.3.3. Other experiments were carried out in a frameless VS under illumination. In a room, 6 × 6 m, two small bars were presented in front of the walls. The walls were covered with black wrinkled plastic sheet material and no corner of the walls was visible. The S looked at the illuminated walls through a rectangular slit and the visual field consisted of only a part of wall vertically extending about 10°. The ceiling and ground were out of sight. A bar consisted of a white rod protruding at a right angle on each side of a yellow circular disk. Four bars of different dimensions were prepared and appropriate ones were used according to the distance e_0 at which these were presented. Bars always were at the eye-level, DP($\theta = 0$), and the S could rotate the bar by a remote control. The head of S was fixed. It will suffice here to explain only two experiments.

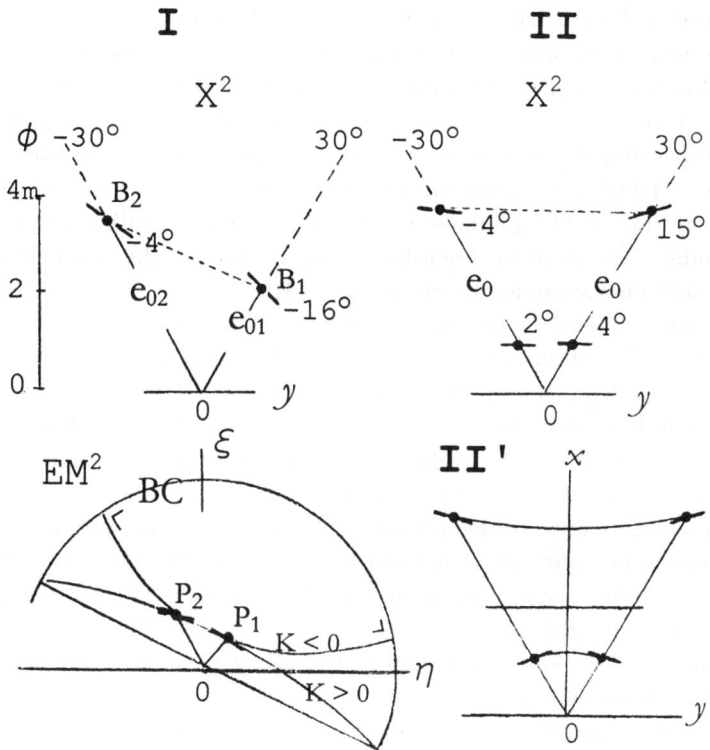

Fig.5.14 Examples of perceptual collinearity settings

One is concerned with visual perception of collinearity (Cuijpers, 2000; Cuijpers, *et al.*, 2002). Various layouts of two bars, B_1 and B_2, were presented and the S was asked to rotate B_1 and B_2 so that "both bars appear to be tangent to an imaginary straight line connecting the center of the bars". Each of four naive S's repeated settings three times in each layout. Fig.5.14I, II are schematic illustrations of settings of a S (RF) in two layouts. The setting of a bar was called veridical when it was oriented along the dotted line connecting the centers of B_1 and B_2. The deviations from veridical setting, $\Delta 1$ in B_1 and $\Delta 2$ in B_2, in degrees with directional sign are included in I and II. If collinearity holds in X^2, $\Delta 1 = \Delta 2 = 0$, which was not the case. Collinearity is a phenomenon in VS^2. If

VS^2 is an R^2, we can think of the geodesic passing through P_1 and P_2 in EM^2 that is either orthogonal to the basic circle BC ($K < 0$) or antipodal ($K > 0$) (Fig.3.1). What is observed in X^2 is the projection of tangents to the geodesic at P_1 and P_2. In order to predict the orientations of B_1 and B_2 in this way, we need the value of K and the mapping functions. In the layout II in which $e_{10} = e_{20} = e_0$, B1 and B_2 must be tangents of a frontoparallel H-curve in X^2. As an experimental fact, we know how H-curve changes its form according to distance. Two examples are shown in Fig.2.1, though the extension in ϕ is much smaller. Irrespective of whether VS^2 is to be regarded as an R^2 or not, the most plausible forms of H-curves are as shown in Fig.5.14II.

Though it is technically difficult to carry out, suppose the following experiment. Two points A and B in the X under discussion are connected by an elastic string and the S is asked to adjust its form so as to appear as the straight line between A and B. When $e_{A0} = e_{B0}$ in Fig.5.14, it is an H-curve. Even when $e_{A0} \neq e_{B0}$, the S will have no difficulty in adjusting the string as instructed. However, it does not guarantee that, when two small portions of both ends of the adjusted curve alone are presented, these are perceived to be collinear. The results of this experiment seem to indicate that it is not the case in general. It does not disprove, however, that the S can see the straight line between A and B. This is the same problem inherent in the bottom-up approach that was mentioned with the experiment of Zimmer.

The other article to be explained here is concerned with visual parallelism (Cuijpers, 2000; Cuijpers, *et al.*, 2000b). Though it consists of two experiments, only one is explained. The experiment was conducted in the same room with the same bars in the experiment on collinearity. Two bars, B_1 and B_2, were presented at the same distance e_0, and B_1 is fixed in the direction ϕ_1 with an orientation angle α_1. Orientation angle α_1 was varied as shown in Fig.5.15I. The other bar B_2 was presented in the direction that is separated from B_1 with an angle ω. The ranges over which e_0, ϕ_1, α_1, and ω were varied are given in Fig.5.15. The task of S was to adjust the orientation angle α_2 so that B_2 appears parallel with B_1. Four S's participated. The number of repetitions of setting by a S was not stated.

The investigators expected that the deviation $\Delta\alpha = \alpha_1 - \alpha_2$ is zero, if the VS^2 under discussion is of constant K. They found "large systematic deviations". Also individual differences were large. Fig.5.1.5II is a schematic illustration of the result of a S, AV, $\Delta\alpha$ at the four angles of separation ω, when $e_0 = 1.47$ m, $\phi_1 = 30°$, and $\alpha_1 = 30°$. The values of $\Delta\alpha$ are given in the figure. There is a systematic change of $\Delta\alpha$. Except when $\alpha_1 = 0°$ and $90°$, a tendency is noticeable that $\Delta\alpha$ is proportional to ω, and when $\alpha_1 = 0°$ or $90°$, $\Delta\alpha$ is close to zero.

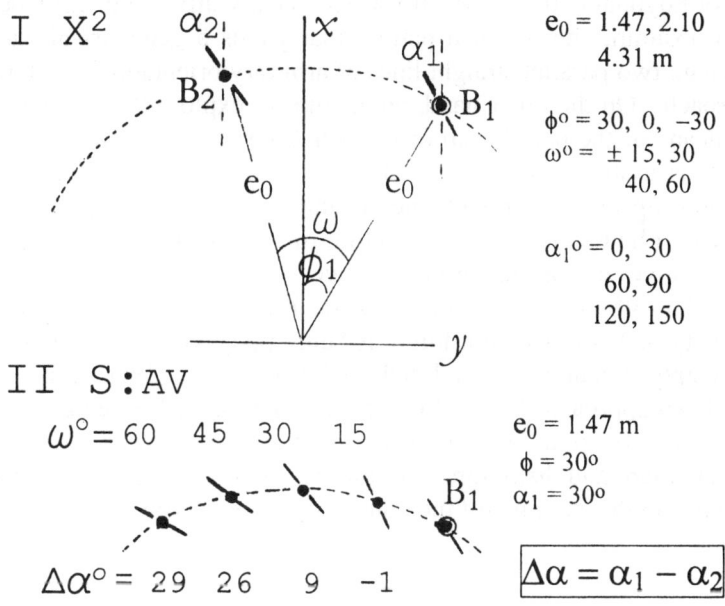

Fig.5.15 Examples of perceptual parallelism settings

To set B_2 at ω from B_1 to appear parallel to B_1 may be regarded as the displacement of a small vector along a curve C(s) in the sense of Levi-Cività. The vector has an angle A with C(s) at $\omega = 0$ and keeps that angle while moving along C(s). The curve needs not be a geodesic. Because B_1 and B_2 appearing parallel is a phenomenon in VS, this displacement must be considered in VS or in EM. The S sees two bars only at the same distance against the wrinkled black background and the dotted curve in Fig.5.15 did not exit in VS. It is not clear how the S judges the parallelism between B_1 and B_2 and how orientation angle α in X is mapped to EM under this condition. When $\alpha_1 = 90°$, $\Delta\alpha$ is close to zero in all values of e_0 and ϕ_1 for three out of the four S's. It may be due to the circumstances that B_1 is close to the tangent to the H-curve in X^2. When $\alpha_1 = 0°$, $\Delta\alpha$ is close to zero in all values of e_0 and ϕ_1 for all the four S's. In this case, B_1 is close to the tangent to the P-alley curve in X^2 that is orthogonal to the η-axis. In other orientation angles α_1, B_1 and B_2 are two small segments parallel in an oblique direction. Once, I tried to construct an obliquely oriented P-alley in X^2 on the DP($\theta = 0$), instead of one orthogonal to the η-axis. It was technically difficult to carry out.

To examine the physical pattern that yields a geometrical pattern in VS, *e.g.*, two parallel straight lines of arbitrary orientation, is a top-down approach. On the other hand, an attempt to explore the structure of VS by means of the parallelism of two small bars is a bottom-up approach. If VS is a solid container in which two small bars satisfying a given relationship or two straight lines forming a geometrical pattern can be placed without changing the structure of the container, both approaches would give consistent results. However, as emphasized in VS5 in Sec.1.1.1, VS is dynamic and what we must explore is the geometrical property of VS in which a figure is being perceived, not the structure of the empty container. Koenderink and van Doorn (1998) criticized the top-down approach starting from the assumption of VS being an R that it is like an attempt to build a house from the roof. However, the house is already there and disjointing it to pieces is not necessarily a good way to understand the structure of the house as a united entity.

5.4. Discussion on the Postulate that VS is an R

Since non-Euclidean geometries have acquired citizenship in science, a new problem has emerged. Which geometry governs the physical space X ? Discussion about this problem is a good guide to the question of which geometry is most appropriate to describe VS.

5.4.1. *Helmholtz-Lie Problem*

A solid physical body can move in the physical space X without deformation. In 1868, Helmholtz argued that such free-mobility could occur only in a very few Riemannian spaces and claimed to have proved that X must be of constant curvature K, either Euclidean (K = 0), or hyperbolic (K < 0) or spherical (⊃elliptic) (K > 0). Eighteen years later, Lie noticed in the argument of Helmholtz a tacit substitution of infinitesimal free-mobility for free-mobility in the large. Adhering to the traditional differentiability assumption, Lie showed that free-mobility of triples of points in a space of less than four dimensions is possible only when the space is of constant K. Hence, discussion on the space that allows free-mobility is called the *Helmholtz-Lie* problem (Freudenthal,1965; Suppes, *et.al*, 1989, *etc.*).

Busemann (1942, 1955, 1959) discussed the Helmholtz-Lie problem without touching upon differentiability of space. For our purpose to characterize the geometrical property of VS, this formulation is more useful because we have no way to prove differentiability of VS. Instead of Riemannian space, Busemann defined G-space, where the G alludes to geodesic that has all the properties apart from differentiability.

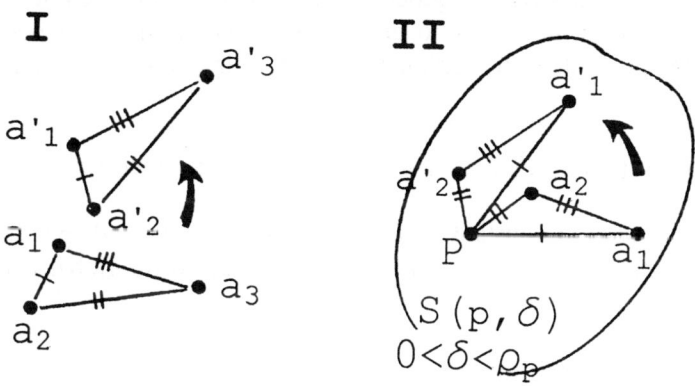

Fig.5.16 Motions in G-space

The axioms for a G-space

G1. The space is metric. Namely, we can define distance $d(x, y)$ between any two points x and y that satisfies the Fréchet conditions (Sec.2.2.1).

G2. The space is finitely compact. Namely, a bounded infinite set has an accumulation point.

G3. The space is convex. Namely, for any two distinct points x, z, a point y with the connection (x, y, z) exists, which means $d(x, z) = d(x, y) + d(y, z)$.

G4. Denote by $S(p, \rho)$ the set of x with $d(p, x) < \rho$. Every point p has a neighborhood $S(p, \rho_p)$, $\rho_p > 0$, such that for any two distinct points x, y in $S(p, \rho_p)$, a point z with (x, y, z) exists.

The four axioms yield geodesics. Namely, any two points x, y can be connected by a segment $T(x, y)$, i.e., a set isometric to an interval of the real t-axis. $T(x, y)$ can be represented in the form of $z(t)$, $\alpha \le t \le \beta = \alpha + d(x, y)$ with

$$d(z(t_1), z(t_2)) = |t_1 - t_2|$$

and $z(\alpha) = x$, $z(\beta) = y$. [α, β] can be extended to all real t, so that it represents a geodesic that is equivalent to a straight line. Busemann considers the other type of geodesic equivalent to a great circle. In this context, it is not necessary to discuss this second type of geodesic. In order to guarantee the uniqueness properties, one more axiom is necessary.

G5. If (x, y, z_1), (x, y, z_2) and $d(y, z_1) = d(y, z_2)$, then $z_1 = z_2$.

Three spaces of constant K, Euclidean, hyperbolic, and spherical, are called elementary, and Busemann gave the following global and local answers to the Helmholtz-Lie problem. Suppose a G-space, X.

Suppose isometric triples, a_1, a_2, a_3 and a'_1, a'_2, a'_3 in X. If a motion of X exists taking a_i into a'_i, ($i = 1, 2, 3$) (Fig.5.16I), then X is elementary.

Suppose every point p in X has a neighborhood $S(p, \delta)$, $0 < \delta < \rho_p$, such that for any four points a_1, a_2, a'_1, a'_2 in $S(p, \rho_p)$ with $d(p, a_i) = d(p, a'_i)$, $i = 1, 2$, $d(a_1, a_2) = d(a'_1, a'_2)$. If a motion of $S(p, \delta)$ exists that takes a_i into a'_i, $i = 1,2$ (Fig.5.16II), then the universal covering space of X is elementary.

Busemann also gave an answer to the inverse problems of the calculus variation. The geodesic between $P(t_1)$ and $P(t_2)$ is obtained by integrating line elements along the shortest path between these points (Eq.2.4.5). This calculus variation is a bottom-up definition of geodesic. The inverse problem is a top-down definition. Given a set of curves in X. It can be decided whether they occur as the extremals of a variational problem.

The global answer to the Helmholtz-Lie problem means that in order to allow free-mobility of a triangle in a G-space X, X must be of constant curvature K. Wang (1951, 1952) took one step forward.

If X^m is two-point homogeneous and m ≥ 2 and odd, then X is elementary.

What is meant by two-point homogeneity is that for any isometric two pairs, a_1, a_2, and a'_1, a'_2, with $d(a_1, a_2) = d(a'_1, a'_2)$, a motion exists taking a_i to a'_i, $i = 1, 2$.

5.4.2. Congruence and Similarity in VS

When we drag a figure or a bar on a CRT monitor as X^2, the figure or the bar moves without deformation and it appears in that way in our VS (veridicality of VS in the neighborhood of the self, VS4 in Sec.1.1). The objective free-mobility holds because the flat surface of the monitor is E^2. The perceptual free-mobility implies that this perceived plane not extending in the direction of depth is of constant K. When we enlarge a figure on the monitor, the perceived figure is also enlarged without distortion. Namely, we see a similarity between the original figure and the enlarged one. This fact implies that K = 0. The perceived display is a small part in VS^3 and its local Euclideanity was regarded as a basis to assume VS to be Riemannian (Sec.2.4.1). Now let us consider the geometrical character of more extended parts in VS.

In a larger part of VS, it is not easy to demonstrate continuous movement of a solid percept. The only exceptional case is construction of D-alley by moving points. Cases on the frameless horizontal DP are shown in Fig.2.2A,B where the S adjusts $Q(x_i, \pm y_i, 0)$ so that the apparent movement of two points $Q(x_i, y_i, 0)$ and $Q(x_i, -y_i, 0)$ along the x-axis appears to keep the same separation. The experiment was performed with light points in the dark or small objects in the illuminated but frameless space. This is free-mobility of a fixed interval in DP. Construction of D-alley in the horizontal direction on a frontoparallel plane HP is shown in Fig.3.4. The HP passed through a fixed point M3 at about $x = 3$ m, and the S adjusted remaining light points Q's so that three series, U, M, L, appeared horizontally parallel (Ph-alley) or vertically equi-distant (Dh-alley). It was found that $\{Q_i\}_P$ for Ph-alley and $\{Q_i\}_D$ for Dh-alley perfectly coincided. Fig.3.5 is an example in which configuration $\{Q_i\}$ of stationary points was adjusted by a S. Two other S's showed the same coincidence between Ph- and Dh-alleys with stationary points Q_i. These two S's participated in another experiment to construct Ph-alley and Dh-alley with three moving points Q_U, Q_M, Q_L (Fig.1, (b) and Table 1 in Indow and Watanabe, 1988). Three light points appeared to move horizontally on the HP. Again, $\{Q_i\}_P$ and $\{Q_i\}_D$ were very close. In the experiments to construct D-alleys by moving points, the S perceives the movement of points P_V's keeping the interval constant. This is free-mobility of a fixed interval in HP.

In contrast to dragging a figure on a monitor, the physical interval between Q's must be adjusted as shown in the Figures to keep perceptual size constant. However, the free-mobility of P_V's and the movement of Q's are related through a set of general mapping functions that are not bounded to this movement. Hence, it is consistent with the Wong's axiom (two-point homogeneity in VS) to postulate that these planes in VS, DP and HP, are respectively of constant K. More convincing evidence for the postulate is afforded by the fact that we can see congruence and/or similarity between figures in these subspaces of VS. When two figures are the same in shape and size, the two are called *congruent*. By moving one into the other, we can superimpose one on the other. When two figures are the same in shape but different in size, the two are called *similar*. In similarity, side lengths are different but ratios between side lengths and angles are the same between the two figures. Strictly speaking, congruence holds on a plane when it is of constant K, and similarity holds only on a flat plane where $K = 0$.

Let us consider the following Gedanken experiment. Suppose there are two triangles on the HP of Fig.5.17 and the right one is the standard. Each triangle can be three light points in the dark, or one having continuous sides on a flat board under illumination. Then, the S can adjust the left one, no matter where it is, so as to appear congruent or similar to the standard. Differing from dragging a figure on the CRT monitor, the two congruent triangles may not be physically the same when they are far apart on the HP. Suppose there are two triangles on the DP of Fig. 5.17 and the right one is the standard. Each triangle can be three light points in the dark or one having continuous sides on a flat board under illumination. Then, the S may be able to adjust the left one, no matter where it is, so as to appear congruent to the standard. In this case, it is very likely that the two triangles are physically not the same.

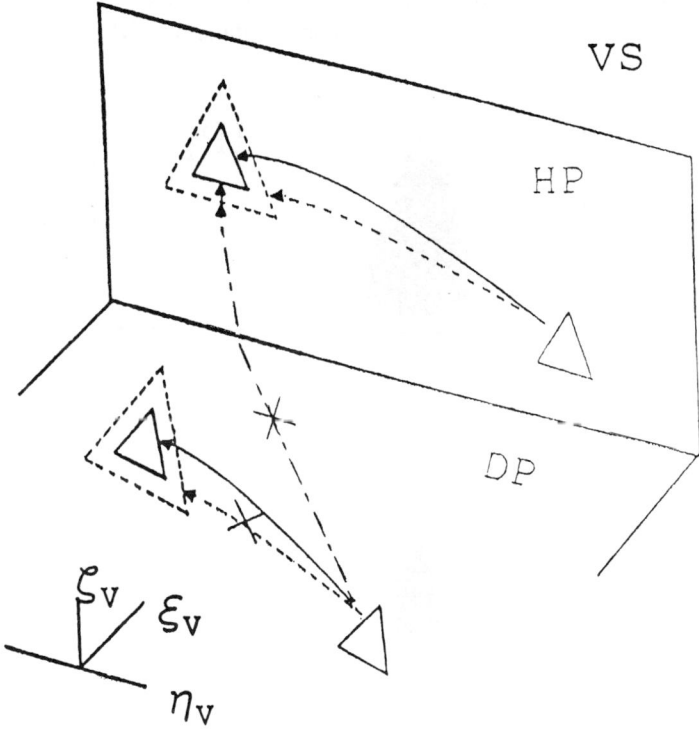

Fig.5.17 Congruence and similarity in VS^3

However, when asked to adjust the left one so as to appear similar to the standard, the S will experience difficulty when the two are sufficiently different in size. If all side lengths are adjusted to be in proportion, some angles may not appear the same. Furthermore, suppose a case in which the standard triangle is on the DP and the other triangle is on the HP. In this case, the S will have difficulty in adjusting the latter so as to appear exactly the same with the standard. It is impossible to imagine a continuous move of any perceived figure from DP to HP, or *vice versa*, that goes through its borderline (the broken curve in Fig.5.17). If this conjecture is correct, it is consistent with the assertion that $K = 0$ in a HP and $K < 0$ in the DP. We can imagine a continuous and shape preserving move of a figure within a slanted plane ($\vartheta_V \neq 0$), because a slanted plane is shown to be of constant K ($\neq 0$). If we imagine a continuous move of a perceived figure between DP and HP through a slanted plane, the same problem occurs at the borderline between the slanted plane and DP as well as at the borderline between the slanted plane and HP.

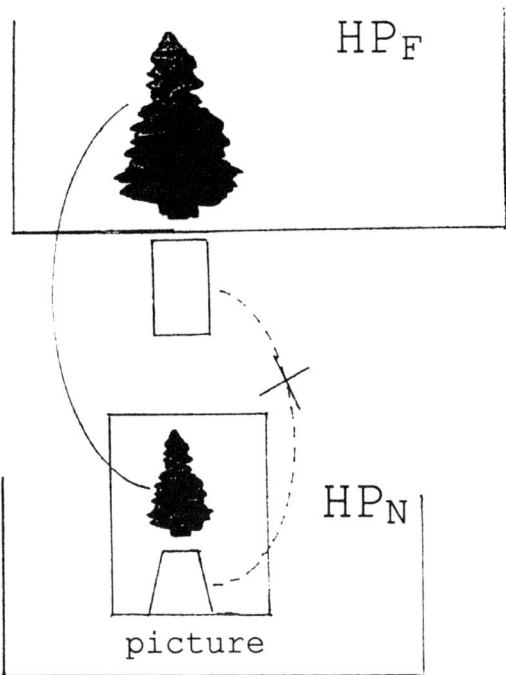

Fig.5.18 Relationship between objects and their images in a picture

Fig.5.18 shows the relationship between a tree on an HP_F at a distance and its picture at hand. For the sake of simplicity, let us assume that the picture is observed perpendicularly to the line of sight. Namely, the picture is on an HP_N. Then, physically and perceptually, the tree in the picture is similar to the tree on HP_F. It is possible because both HP_F and HP_N are of $K = 0$. Suppose there is a rectangular thin object on the ground under the tree. Its image in the picture is not rectangular. It is a problem of perspective how to make its appearance natural to human eyes.

The range that our attention can focus on at a time is limited. When we are paying attention to the side lengths, perhaps we do not care about small deviations in angles between two figures. Hence, we may have tolerance of a considerable degree in perceiving congruence or similarity. What is discussed above is certainly oversimplified. Nevertheless, the discussion will be illustrative of how perception of congruence or similarity is related to the geometrical relations between two perceived figures, the original one and its replica. As mentioned in Foreword, modern technology has made it possible to adjust a displayed pattern in a much more flexible way than ever before, and it may not be a dream anymore to materialize the Gedunken experiments described with regard to Figs.5.17, 18.

5.4.3. *Linear Perspective*

The scene in the Fig.5.18 consists of a flat tree only. When the scene includes other objects at different distances and each object is a solid form, it is the problem of linear perspective how to draw the scene on a flat picture plane HP. In Fig.5.18, this problem occurred with the rectangular object on the ground. Linear perspective is "invention" by Filippo Brunelleschi (1377-1446), Leon Battista Alberti (1404-1472), and Leonardo da Vinci (1452-1519), *etc*. Pictures before the Renaissance look very flat (*e.g.*, Byzantine arts). Linear perspective is a set of rules based on mathematical formulation and called "the science of art" (Kemp, 1990).

Suppose a case that the square grid on the ground viewed by a person at G_0 (Fig.5.19A) is drawn on the picture plane. The position of the eyes is denoted as O, and the picture plane is frontoparallel (an HP). The

194 *Global Structure of Visual Space*

central pattern in Fig.5.19B shows the picture drawn on the HP, which is obtained by the intersections with the HP of lines connecting points on the grid lines and O. Lines in the grid, 1 to 5, that are supposed to extend to infinity, are represented as lines that converge to a point F at the height of eyes. This is called the vanishing point V in one-point perspective. The pattern of these lines is independent of the distance (G_0 G_F). The line that passes through V and is parallel with the ground is the horizontal line. Lines in the grid, I to IV, that are parallel to the base line,

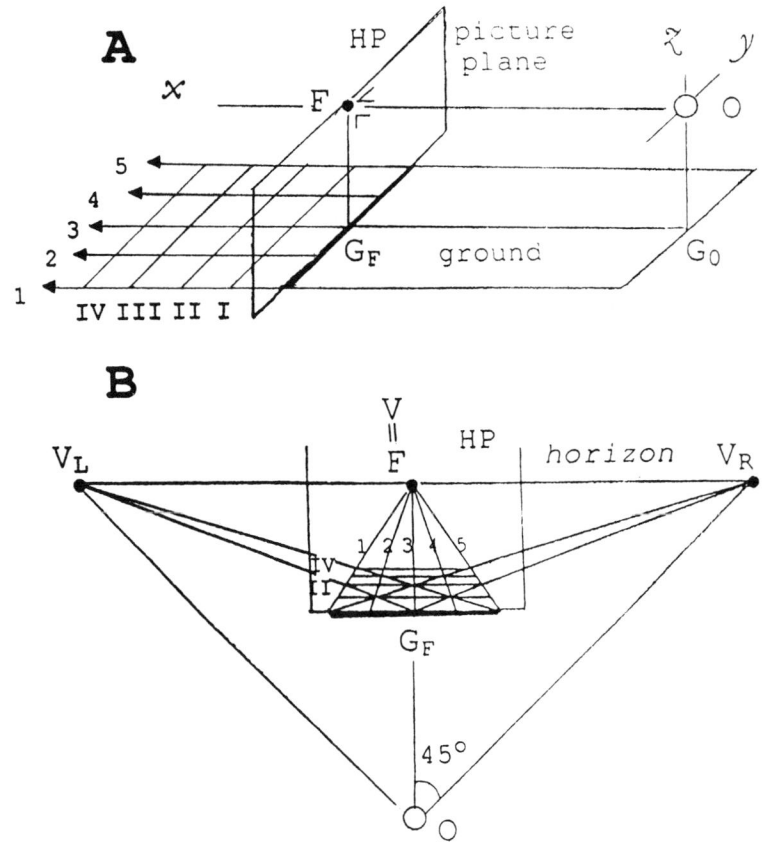

Fig.5.19 Two-point perspective

are represented as shown in B. These lines change according to the distance (G_0 G_F). The square grid becomes receding rectangles having the following property. All diagonal lines through the rectangles converge to a point, either V_R or V_L, on the horizon line, where (V, V_R) = (V, V_L) = (F, O). In other words, $\angle VOV_R = \angle VOV_L = 45°$. Often V_R and V_L are called lateral vanishing points or distance points in two-point perspective. When a solid object to be drawn is tilted with regard to the z-axis, we need three vanishing points, VL, VR, and VU in Fig.5.20 (three-point perspective).

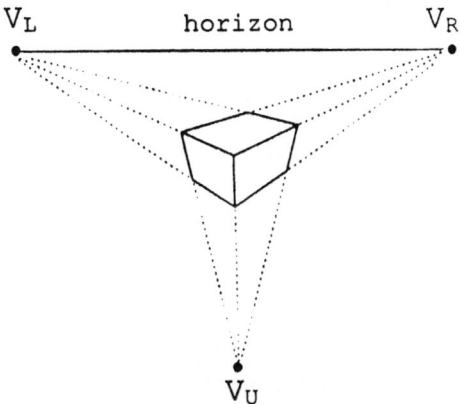

Fig.5.20 Three-point perspective

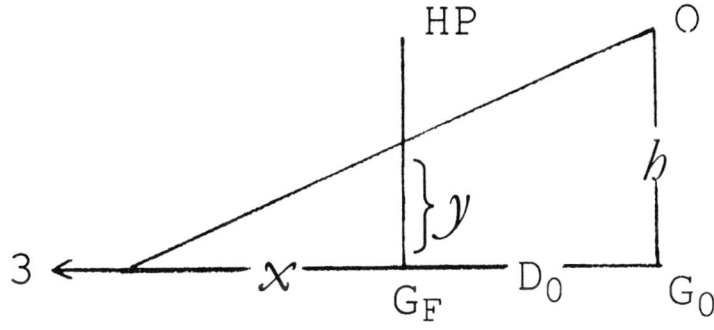

Fig.5.21 The vertical plane on the line 3 in Fig.5.20

These rules are based on Euclidean projective geometry and trigonometry in E^3. Take the vertical plane on the central line 3 and G_0, G_F in Fig.5.19A and denote the distance of a point from G_F as x, the distance from G_F to the projected point on the HP as y, and the height of O as h (Fig.5.21). Then,

$$y = h \frac{x}{D_0 + x} \qquad (5.4.1)$$

$$\frac{dy}{dx} = \frac{h}{D_0} \frac{1}{(1 + \frac{x}{D_0})^2} \qquad (5.4.2)$$

These equations show how lines I, II, *etc.* on the ground in Fig.5.19A are projected to HP in Fig 5.19B. It is interesting to see that Eq.(5.4.1) is the same with the distance function of Gilinsky (Eq.4.2.2) if y is replaced by d and D_0 by M. Obviously, h = max d.

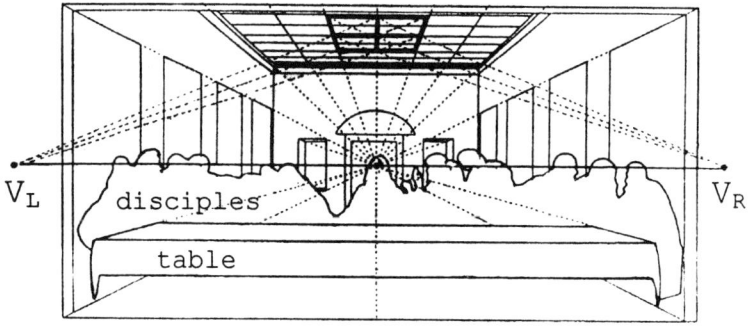

Fig.5.22 Perspective of the Last Supper

Fig.5.22 shows the two-point perspective of the Last Supper by Leonardo da Vinci based on the analysis in pl.80 in Kemp (1990). Christ and disciples at the table are shown in silhouette. The vanishing point V is on the eyes of Christ in the center. The painting is placed high in the refectory of Santa Maria delle Grazie. Hence, the correct viewing position, O in Fig.5.19, is about 15 ft above the floor. The ceiling in the painting is six coffers wide and twelve coffers deep. Hence, two adjacent coffers in the depth direction are put together to have a square. Then, diagonals converge to distant points V_R and V_L on the horizon line (only three diagonals are shown for each distant point). Obviously, Fig. 5.21 is an oversimplified analysis of perspective, and Kemp wrote "closer analysis reveals a series of ambiguities and artifices which save the appearance of optical legitimacy while acknowledging the inbuilt problems and contradictions of perspectival illusion in a given situation" (p.47).

In a different context, Kemp emphasizes that a geometrical understanding of the technical procedures of perspective needs to be seasoned by experience of pictorial practice in representing scenes (p.228). The necessity of seasoning may be ascribed, partially at least, to the circumstances that construction of scene based on Euclidean geometry is not quite appropriate for human perception of the scene. Finch (1977) pointed out that two-point perspective of the square grid on the ground gives rise to distortions in the receding squares on the far right and left sides near the horizon. According to him, these distortions can be reduced if the grid is drawn on a hyperbolic surface.

5.4.4. Concluding Remarks

Throughout this book, stable VP consisting of stationary patterns is dealt with. It is generated through multiple glances of a stationary spatial layout in the physical space X. Neither moving patterns nor movement of the observer is involved. Even if alleys were constructed by moving points (Fig.2.2), the alley itself did not move in VS. Since VS is based on multiple glances, the retinal image during the observation is not stationary. In most cases, a flat visual pattern $\{P_{Vi}\}$, either on a DP plane or on a HP plane, is discussed. The corresponding stimulus pattern $\{Q_i\}$ in X is not necessarily flat. In general, a pattern $\{P_{Vi}\}$, flat or solid, is oriented with regard to the self in the VS^3, which means that either the entire VS is frameless or both $\{P_{Vi}\}$ and the self are in the same global framework.

1. The approach of this book is phenomenological at the psychophysical level (Fig.1.1). The physiological processes between the retinal image and VS are left in a black box. Though unknown, we are equipped with mechanism that makes it possible for us to have this highly structured end-product.

2. The main concern is the geometrical structure of $\{P_{Vi}\}$. For the reason stated in Secs. 5.4.1, and 5.4.2, it was tried to understand the structure in terms of Riemannian geometry of constant curvature R. This is a top-down approach. There are other approaches to VS that are of more differential geometry type (*e.g.*, Hoffman, 1966, 1977, 1980; Yamazaki, 1987). To go into these approaches, we need different mathematical preparations.

3. The interpretation of $\{P_{Vi}\}$ in terms of R makes it possible to formulate all equations in EM, and to have more flexibility in cope with the context-dependent correspondence between a pattern $\{P_{Vi}\}$ in EM and the stimulus pattern $\{Q_i\}$ in X.

4. It has been emphasized that VS is dynamic and context-dependent (VS.5 in Sec.1.1.1). VS^3 is not a solid container into which various percepts are placed without affecting the structure of the container. What has been discussed is the geometry of VS^3 with a given configuration of percepts, not the geometry of VS^3 as the general container. Furthermore, when a pattern $\{P_{Vi}\}$ is generated by a physical layout $\{Q_i\}$, the geometrical structure of the $\{P_{Vi}\}$ is determined not only by the $\{Q_i\}$ but also by other conditions in X^3. As stated at the end of Sec.5.2.3, the purpose of this book is not to predict $\{P_{Vi}\}$ when $\{Q_i\}$ is given, but to make explicit the geometrical structure of $\{P_{Vi}\}$ when it is generated.

A number of theoretical articles on VS and spatial representation in general were written by Suppes (*e.g.*, 1977, 1989, 1995, 2002) and Drösler (*e.g.*, 1966, 1979, 1988, 1992, 1994). In the article of 1977, Suppes reviewed studies concerning the question "is visual space Euclidean ?" What is dealt with therein is mainly a frameless VS with various $\{P_{Vi}\}$ and he listed in Table 1 answers of respective studies to this question; 3 "yes", 9 "no", and 4 "not sure". He wrote in the article of 1995 "The goal of having a unified structure of visual space adequate to account for all the important experimental results now seems mistaken. A pluralistic and fragmentary approach seems required" (p.43). The approach taken in this book is to give pluralistic and fragmentary answers to geometrical structure of $\{P_{Vi}\}$. However, all answers are in terms of R. No matter whether VS is frameless or not, when a $\{P_{Vi}\}$ is generated on a HP plane, it is of $K = 0$. When a $\{P_{Vi}\}$ is generated on a DP plane, it is of $K < 0$ and the value of K is context-sensitive. Furthermore, the correspondence between $\{P_{Vi}\}$ and $\{Q_i\}$ is also context-dependent.

Drösler develops equations directly in X. In the article of 1988, in which Vieth-Müller circle and Hillebrand hyperbola (Fig.5.9) were discussed, he wrote "The psychophysical function for binocular vision maps the physical stimuli into a second structure, which is visual space. Sometimes a third structure, the Euclidean map of visual space (cf. Luneburg 1947; Indow, 1979) is mentioned. However, it is of interest only after the intrinsic geometric structure of visual space has been determined" (p.296). In contrast, VS under a fixed physical layout is assumed to be a R in this book and all equations are first formulated in EM. To cope with the dynamic property of VS, the value and even sign of K are assumed to change from a layout to another. Furthermore, the mapping functions between $\{P_{Vi}\}$ and $\{Q_i\}$ are also assumed to change in accordance with the context. The mapping functions need not be in terms (γ, ϕ, θ). As Drösler (1992) once discussed VS with monocular vision, it is not necessary to use the binocular cue γ in formulating the mapping functions. It may be more reasonable to think of three arguments, $e(\phi,\theta)$, and ϕ, θ. In many articles and books, VS is often called binocular visual space. This word has never been used in this book.

References

Aczél J., Boros Z., Heller J. and Tat Ng C. Fundamental equations in binocular space perception, *Journal of Mathematical Psychology* **43**(1999), 71-101.

Amigo G. A vertical horopter, *Optica Acta* **21**(1974), 277-292.

Baird J.C. and Wagner M. The moon illusion: I. How high is the sky ? *Journal of Experimental Psychology; General* **111**(1982), 296-303.

Battro A.M., di Pierro A. and Rozestraten R.J.A. Riemannian geometries of variable curvature in visual space: Visual alleys, horopters, and triangles in big open fields, *Perception* **5**(1976), 9-27.

Blank A.A. The geometry of vision, *British Journal of Physiological Optics* **14**(1957), 1-30.

Blank A.A. Axiomatics of binocular vision. The foundations of metric geometry in relation to space perception, *Journal of Optical Society of America* **48**(1958), 328-334.

Blank A.A. Curvature of binocular visual space, *Journal of Optical Society of America* **51**(1961), 335-339.

Blank A.A. Metric geometry in human binocular perception: Theory and fact, in *Formal Theories of Visual Perception*, eds. Leeuwenberg E.L.J. and Buffart H.F.J.M. (John Wiley, New York, 1978), 83-102.

Blumenfeld W. Untersuchungen über die sheinbarere Grösse in Sheraume, *Zeitschrift für Psychologie* **65**(1913), 241-404.

Busemann H. *Metric Methods in Finsler Spaces and in the Foundations of Geometry* (Princeton University Press, Princeton, 1942).

Busemann H. and Kelly P.J. *Projective Geometry and Projective Metrics* (Academic Press, New York, 1953).

Busemann H. *Geometry of Geodesics* (Academic Press, New York, 1955).

Busemann H. Axioms for geodesics and their implications, in *The Axiomatic Method with Special Reference to Geometry and Physics*, eds. Hopkins L., Suppes P. and Tarski A. (North-Holland, Amsterdam, 1959), 146-159.

Cogen A.I. The relationship between the apparent vertical and the

vertical horopter, *Vision Research* **19**(1979), 655-664.
Collet T. and Harkness L.I.K. Depth vision in animals, in *The Analysis of Visual Behavior*, eds. Ingle D., Goodale M. and Mansfield R. (MIT Press, Cambridge, 1982), 111-176.
Cook M. The judgment of distance on a plane surface, *Perception and Psychophysics* **23**(1978), 85-90.
Cooper M.L. and Pettigrew J.D. A neurophysiological determination of the vertical horopoter in the cat and owl, *Journal of Comparative Neurology* **184**(1979), 1-26.
Cornish V. *Scenery and the Sense of Sight* (Cambridge University Press, Cambridge 1935)
Cornish V. *The Scenery of England; A Study of Harmonious Grouping in Town and Country* (Alexander Maclehose, London, 1937).
Cuijpers R.H. *The Geometry of Visual Space* (Helmholtz Institute, University of Utrecht, 2000).
Cuijpers R.H., Kapper A.M.L. and Koenderink J.J. Investigation of visual space using an exocentric pointing task, *Perception and Psychophysics* **62**(2000a), 1556-1571.
Cuijpers R.H., Kapper A.M.L. and Koenderink J.J. Large systematic deviations in visual parallelism, *Percption* **29**(2000b), 1467-1482.
Cuijpers R.H., Kapper A.M.L. and Koenderink J.J. Visual percetion on collinearity, *Perceptin and Psychophysics* **64**(2002), 392-404.
Cutting J.E. and Vishton P.M. Perceiving layout and knowing distances: The integration relative potency, and contextual use of different information about depth, in *Perception of Space and Motion*, eds. Epstein W. and Roger S. (Academic Press, New York, 1995), 69-117.
Da Silva J.A. Scales for subjective distance in a large open field from the fractionation method: Effects of type of judgment and distance range, *Perceptual and Motor Skill* **55**(1982), 283-288.
Da Silva J.A. and Das Santos R.A. Scaling apparent distance in a large open field: Presence of a standard does not increase the exponent of the power function, *Perceptual and Motor Skill* **55**(1982), 267-274.
Da Silva J.A. Scales for measuring subjective distance in children and adults in a large open field, *Journal of Psychology* **113**(1983), 221-230.

Da Silva J.A., Da Silva C.E., Scaling apparent distance in a large open field: Some new data, *Perceptual and Motor Skill* **56**(1983), 135-138.
Drösler J. Das beidäugige Raumsehen, in *Handbuch der Psychologie*, Band I. ed. Metzger, M. (Hogrefe, Göttingen 1966), 590-615.
Drösler J. Foundations of multi-dimensional metric scaling in Cayley-Klein geometries, *British Journal of Mathematical and Statistical Psychology* **32**(1979), 185-211.
Drösler J. The psychophysical function of binocular space perception, *Journal of Mathematical Psychology* **32**(1988), 285-297.
Drösler J. Eine Untersuchung des perspektivischen Sehens, *Zeitschrift für experimentelle und angewandte Psychologie* **39**(1992), 515-532.
Drösler J. The invariances of Weber's and other laws as determinants of psychophysical structures, in *Geometric Representations of Perceptual Phenomena: Papers in honor of Tarow Indow of his 70th birthday*, eds. Luce D. et al. (Lawrence Erlbaum, Mahwah, N.J., 1995), 69-93.
Dunker K. Über induzierte Bewegung, *Psychologische Forshung* **12**(1929), 180-259.
Dusek E.R., Teichner W.H. and Kobrick J.L. The effects of the angular relationships between the observer and the stimulus surround on relative depth discrimination, *American Journal of Psychology* **68**(1955), 438-443.
Engel W. Optische Untersuchungen am Ganzfeld. I. Die Ganzfeldanordnung, *Psychologische Forchung* **13**(1930), 1-5.
Eschenburg J.-H. Is binocular visual space constantly curved ? *Journal of Mathematical Biology* **9**(1980), 3-22.
Falmagne J-C. *Elements of Psychophysical Theory* (Oxford University Press, New York), 1985.
Fieandt v.K. *The world of Perception* (Dorsey Press, Homewood, IL, 1966),
Filehne W. Die mathematische Ableitung der form des scheinbaren Himmelsgewöbes, *Archiev für Physiologie* **5**(1912), 1-32.
Finch D. Hyperbolic geometry as an alternative to perspective for constructing drawings of visual space, *Perception* **6**(1977), 221-225.
Foley J.M. Visual space: A test of constant curvature hypothesis,

Psychological Science **1**(1964a), 9-10.
Foley J.M. Desarguesian property in visual space, *Journal of Optical Society of America* **54**(1964b), 684-692.
Foley J.M. Locus of perceived equidistance as a function of viewing distance, *Journal of Optical Society of America* **56**(1966), 822-827.
Foley J.M. The size-distance relation and intrinsic geometry of visual space: Implications for processing, *Vision Research* **12**(1972), 323-332.
Foley J.M. Primary distance perception, in *Handbook of Sensory Physiology, Volume VIII Perception*, eds. Held R., Leibowitz H.W. and Teuber H.-L. (Springer, New York, 1978).
Foley J.M. Binocular distance perception, *Psychological Review* **87**(1980), 411-434.
Freudenthal, H. Neue Fassungen des Riemann-Helmholtz-Lieschen Raumproblems, *Mathematische Zeitschrift* **63**(1956), 374-405.
Galanter E. and Galanter P. Range estimates of distant visual stimuli, *Perception and Psychophysics* **14**(1973), 301-306.
Gilinsky A.S. Perceived size and distance in visual space, *Psychological Review* **58**(1951), 460-482.
Gilinsky A.S. The paradoxical moon illusions, *Perceptual and Motor Skill* **50**(1980), 271-283.
Gilinsky A.S. The moon illusion in a united theory of visual space, in *The Moon Illusion*, ed. Hershenson M. (Lawrence Erlbaum, Hillsdale, N.J., 1989), 167-192.
Hagino G. and Yoshioka I. A new method for determining the personal constants in the Luneburg theory of binocular visual space, *Perception and Psychophysics* **19**(1976), 499-509.
Harber R.N. Toward a theory of the perceived spatial layout of scenes, *Computer Vision, Graphics, and Image Processing* **31**(1985), 282-321.
Hardy L.H., Rand G. and Rittler M.C. Investigation of visual space: The Blumenfeld alley, *Archives of Ophthalmology* **45**(1951), 53-63.
Hardy L.H., Rand G. Rittler M.C. and Boeder P. *The Geometry of Binocular Space Perception* (Knapp Memorial Laboratories, Institute of Ophthalmology, Columbia University, New York, 1953).

Harkness L. Chameleons use accommodation cues to judge distance, *Nature* **267**(1977), 346-349.
Harway N.I. Judgment of distance in children and adults, *Journal of Experimental Psychology* **65**(1963), 385-390.
Heelan P.A. *Space-perception and Philosophy of Science* (University of California Press, Berkeley, 1983).
Heller J. On the psychophysics of binocular space perception, *Journal of Mathematical Psychology* **41**(1997), 29-43.
Hershenson M. ed. *The Moon Illusion* (Lawrence Erlbaum, Hillsdale, N.J., 1989), 167-192.
Higashiyama A. Variation of curvature in binocular visual space estimated by triangle method, *Vision Research* **21** (1981), 925-933.
Higashiyama A. Curvature of binocular visual space, *Vision Research* **24**(1984), 1713-1718.
Hillebrand F. Theorie des scheinbaren Grösse bei binocularem Sehen, *Denkschriften der Wiener Akademie, Mathematisch-Naturwissenschaft Klasse* **72**(1902), 255-307.
Hoffman W.C. The Lie algebra of visual perception, *Journal of Mathematical Psychology* **3**(1966), 65-98.
Hoffman W.C. The Lie transformation group approach to visual neuropsychology, in *Formal Theories of Visual Perception*, eds. Leeuwenberg E.L.J. and Buffart H.F.H.M. (John Wiley, New York 1978), 66-77.
Hoffman W.C. Subjective geometry and geometric psychology, *Mathematical* M*odelling* **1**(1980), 349-367.
Howard I.P. and Roger R.J. *Binocular Vision and Stereopsis* (Oxford University Press, New York, 1995).
Indow T., Inoue E. and Matsushima K. An experimental study of the Luneburg theory of binocular space perception (1) The 3-and 4-point experiments, *Japanese Psychological Research* **4**(1962a), 6-16.
Indow T., Inoue E. and Matsushima K. An experimental study of the Luneburg theory of binocular space perception (2) The alley experiments, *Japanese Psychological Research* **4**(1962b), 17-24.
Indow T., Inoue E. and Matsushima K. An experimental study of the

Luneburg theory of binocular space perception (3) The experiment in a spacious field, *Japanese Psychological Research* 5(1963), 1-27.

Indow T. Multidimensional mapping of visual space with real and simulated stars, *Perception and Psychophysics* 3(1968), 45-64.

Indow T. Applications of multidimensional scaling in perception, in *Handbook of Perception, Volume II, Psychophysical Judgment and Measurement*, eds. Carterette, E.C. and Friedman, M.P. (Academic Press, 1974), 493-525.

Indow T. and Ida M. On scaling from incomplete paired comparison matrix, *Japanese Psychological Research* 17(1975), 98-105.

Indow T. and Ida M. Scaling of dot numerosity, *Perception and Psychophysics* 22(1977), 265-276.

Indow T. Alleys in visual space, *Journal of Mathematical Psychology* 19(1979), 221-258.

Indow T. An approach to geometry of visual space with no *a priori* mapping functions, *Journal of Mathematical Psychology* 26(1982), 204-236.

Indow T. and Watanabe T. Parallel- and distance-alleys with moving points in the horizontal plane, *Perception and Psychophysics* 35(1984a), 144-154.

Indow T. and Watanabe T. Parallel- and distance-alleys on horopter plane in the dark, *Perception* 13(1984b), 165-182.

Indow T. Alleys on apparent frontoparallel plane, *Journal of Mathematical Psychology* 32(1988), 259-284.

Indow T. and Watanabe T. Alleys on an extensive frontoparallel plane: A second experiment, *Perception* 17(1988), 647-666.

Indow T. On geometrical analysis of global structure of visual space, in *Psychophysical Explorations of Mental Structures*, ed. Gesssler H.-G. (Hogrefe and Huber, New York, 1990), 172- 180.

Indow T. A critical review of Luneburg's model with regard to global structure of visual space, *Psychological Review* 98(1991), 430-453.

Indow T. Hyperbolic representation of global structure of visual space, *Journal of Mathematical Psychology* 41(1997), 89-98.

Indow T. Global structure of visual space as a united entity, *Mathematical Social Sciences* 38(1999), 377-392.

Kahl R. ed. *Selected writings of Hermann von Helmholtz* (Wesleyan

University Press, Middletown, Conn. 1971).
Kaufman L. and Rock I. The moon illusion thirty years later, in *The Moon Illusion*, ed. Hershenson, M. (Lawrence Erlbaum, Hillsdale, N.J., 1989), 193-234.
Kemp M. *The Science of Art* (Yale University Press, New Haven, 1990).
Koenderink J.J. and Doorn v.A.J. Exocentric pointing, in *Vision and Action*, eds. Harris L.R. and Jankin M. (Cambridge University Press, Cambridge, 1998), 295-313.
Koenderink J.J., Doorn v.A.J. and Lappin J.S. Direct measurement of the curvature of visual space, *Perception* **29**(2000), 69-79.
Köhler W. Ein altes Scheinproblem, *Naturwissenschaften* **17**(1929), 395-401.
Krantz D.H. Luce, R.D. Suppes, P. and Tversky, A. *Foundations of Measurement, Volume I. Additive and Polynomial Representations* (Academic Press, New York, 1971).
Krekling S. and Blika S. Development of the tilted vertical horopter, *Perception and Psychophysics* **34**(1983), 491-493.
Kuroda T. Distance constancy: Functional relationship between apparent distance and physical distance, *Psychologische Forschung* **34**(1971), 199-219.
Levin C.A. and Haber P.N. Visual angle as a determinant of perceived interobject distance, *Perception and Psychophysics* **54**(1993), 250-259.
Luce R.D. On the possible psychophysical laws, *Psychological Review* **66**(1959), 81-95.
Lukas J. Visualle Frontalparallelen: Ein Entscheidungsexperiment zu den Theorie von Blank, Foley und Luneburg, *Zeitschrift für Experimentelle und Angewandte Psychologie* **30**(1983), 610-627.
Luneburg R.K. *Mathematical Analysis of Binocular Vision* (Princeton University Press, Princeton, 1947).
Luneburg R.K. The metric of binocular visual space, *Journal of Optical Society of America* **50**(1950), 637-642.
Metzger W. Optische Untersuchungen am Ganzfeld. II. Zur Phänomenologie des homogenen Ganzfeld. *Psychologische Forchung* **13**(1930), 6-29.
Miller A. and Neuberger H. Investigation into the apparent shape of the

sky, *Bulletin of American Meteorological Society* **26**(1945), 212-216.

Nakayama K. Geometric and physiological aspects of depth perception, *Proceedings of Society of Photo-Optical Instrument Engineers* **120**(1977), 2-9.

Neuberger H. General meteorological optics, in *Compendium of Meteorology*, ed. Malone T.H. (American Meteorological Society, Boston, 1951), 61-78.

Ogle K.N. *Researches in Binocular Vision* (Hafner, New York, 1964).

Plug C. and Ross H.E. Historical Review, in *The Moon Illusion*, ed. Hershenson M. (Lawrence Erlbaum, Hillsdale, N.J., 1989), 5-27.

Purdy J. and Gibson E. Distance judgment by the method of fractionation, *Journal of Experimental Psychology* **50**(1955), 374-380.

Ross H.E. and Ross G.M. Did Ptolemy understand the moon illusion ? *Perception* **5**(1976), 377-386.

Schiffman S.S., Reynolds M.L. and Young F.W. *Introduction to Multidimensional Scaling* (Academic Press, New York, 1981)

Shipley T. Convergence function in binocular visual space. II Experimental report, *Journal of Optical Society of America* **47**(1957), 804-821.

Squires P.C. Luneburg theory of visual geodesics in binocular space perception, *Archives of Ophthalmology* **56**(1956), 288-297.

Stevens S.S. Mathematics, measurement, and psychophysics, in *Handbook of Experimental Psychology*, ed. Stevens S.S. (Wiley, New York, 1951).

Stevens S.S. The direct estimation of sensory magnitude-loudness, *American Journal of Psychology* **69**(1956), 1-25.

Stevens S.S. On the psychophysical law, *Psychological Review* **64**(1957), 153-181.

Stevens S.S. *Psychophysics,* ed. Stevens G. (Wiley, New York, 1975).

Suppes P. Is visual space Euclidean ?, *Synthese* **35**(1977), 397-421.

Suppes P., Krantz, D.M., Luce, R.D., and Tversky, A. *Foundations of Measurement Volume II. Geometrical, Threshold, and Probabilistic Representations* (Academic Press, New York, 1989).

Suppes P. Some fundamental problems in the theory of visual space, in *Geometric Representations of Perceptual Phenomena: Papers in honor of Tarow Indow of his 70th birthday*, eds. Luce D. et al. (Lawrence Erlbaum, Mahwah, N.J., 1995), 37-45.

Suppes P. *Representation and Invariance of Scientific Structure* (CSLI Publications, Stanford, 2002)

Techtsoonian M. and Techtsoonian R. Scaling apparent distance in natural indoor settings, *Psychonomic Science* **16**(1969), 281-283.

Techtsoonian R. and Techtsoonian M. Scaling apparent distance in a natural outdoor setting, *Psychonomic Science* **21**(1970), 215-216.

Teichner W.H., Kobrick J.L. and Wehrkamp R.F. The effects of terrain and observation distance on relative depth discrimination, *American Journal of Psychology* **68**(1955), 193-208.

Torgerson W.S. Multidimensional scaling, I. Theory and method, *Psychometrika* **17**(1952), 401-419.

Toye R.C. The effect of viewing position on the perceived layout of scenes, *Perception and Psychophysics* **40**(1986), 85-92.

Tyler C.W. Cyclopean vision, in *Binocular Vision*, ed. Regan, D. (CRC Press, Boca Raton, 1991), 38-74.

Wagner M. The metric of visual space, *Perception and psychophysics* **38**(1985), 483-495.

Wang H.C. Two theorems on metric spaces, *Pacific Journal of Mathematics* **1**(1951), 473-480.

Wang H.C. Two-point homogeneous spaces, *Annals of Mathematics* **55**(1952), 177-191.

Watanabe T. The estimation of the curvature of visual space with a visual triangle (in Japanese), *Japanese Journal of Psychology* **67**(1996), 278-284.

Yamazaki T. Non-Riemannian approach to geometry of visual space: An approach of affinely connected geometry to visual alleys and horopter, *Journal of Mathematical Psychology* **31**(1987), 270-298.

Zajaczkowska A. Experimental determination of Luneburg's constants σ and K, *Quarterly Journal of Experimental Psychology* **8**(1956a), 66-78.

Zajaczkowska A. Experimental test of Luneburg's theory: Horopter and alley experiments, *Journal of Optical Society of America* **46**(1956b), 514-527.

Zimmer K. *Experimentelle Untersuchungen zur geomtrischen Struktur des binokularen Sehraums* (Shaker Verlag, Aachen, 1998).

Author Index

Aczél, A. 173
Amigo, G. 12
Baird, J.C. 112
Battro, A.M. 40
Blank, A.A. 36, 152, 159, 163, 175
Blika, S. 12
Blumenfeld, W. 19
Busemann, H, 53, 187
Cogen, A.I. 12
Collet, T. 15
Cook, M. 133
Cooper, M.L. 14
Cornish, V. 119
Cuijpers, R.H. 183, 184,
Cutting, J.E. 144
Da Silya, C.E. 132
Da Silva, J.M. 132, 133, 135
Das Santos, R.A. 132
Drösler, J. 7, 9, 198, 199
Doorn, v.A.J. 145, 146, 186
Dunker, K. 3
Dusek, R.R. 126
Engel, W. 6
Eschenburg, J. –H. 36, 171
Falmagne, J.-C. 181
Fieandt, v. K. 5
Filehne, W. 104, 107
Finch, D. 197
Foley, J.M. 159, 164, 165, 166, 167, 176, 177, 179
Freudenthal, H. 187
Galanter, E. 130, 133
Galanter, P. 130, 133
Gibson, E. 133
Gilinsky, A.S. 112, 113, 122, 124, 125, 126, 135, 176, 196
Haber, P.N. 137, 138, 139, 140, 143, 144
Hagino, G. 149, 151

Hardy, L.H. 19, 36, 40, 41
Harkness, L.I.K. 15
Harway, N.I. 123, 135
Heelan, P.A. 144
Heller, J. 163. 165. 171. 174
Helmholtz, v.H. 13
Hershenson, M. 110
Higashiyama, A. 157, 159, 166
Hillebrand, F. 19
Hoffman, W.C. 198
Howard, I.P. 15
Ida, M. 84, 136
Indow, T. 7, 19, 20, 21, 39, 40, 41, 42, 43, 45, 69, 74, 75, 77, 79, 83, 84, 89, 94, 95, 114, 117, 121, 136, 154, 190, 199
Inoue, E. 19, 39, 43
Kahl, R. 14
Kaufman, L. 111
Kelly, P.J. 53
Kemp, M. 193, 196
Koenderink, J.J. 145, 146, 147, 182, 186
Köhler, W. 3
Krants, D.H. 133
Krekling, S. 12
Kuroda, T. 125, 126, 135
Lappin, J.S. 145
Levin, D.A. 139, 140, 143, 144
Luce, R.D. 134
Lukas, J. 163, 165
Luneburg, R.K. 17, 22, 25, 29, 31, 61, 70, 147, 172, 173, 174, 199
Matsushima, K. 19, 39, 43
Metzger, W. 5
Miller, A. 105
Nakayama, K. 12
Neuberger, H. 104, 105, 107

Ogle, K.N. 12, 39, 165
Pettigrew, J.D. 14
Plug, C. 110, 112
Rand, G. 19, 40, 41
Rittler, M.C. 19, 40, 41
Rock, I. 111
Roger, R.J. 15
Ross, G.M. 110
Ross, H.E. 110, 112
Purdy, J. 133
Schiffman, S.S. 80
Shipley, T. 19
Squires, P.C. 19
Stevens, S.S. 129, 133, 167
Suppes, P. 180, 187, 198
Techtsoonian, M. 132
Techtsoonian, R. 132

Teichner, W.H. 126, 127, 135
Torgerson, W.S. 87
Toye, R.C. 137, 139, 142, 143, 144
Tyler, C.W. 12
Vishton, P.M. 144
Wang, H.C. 189
Wagner, M. 112, 140, 142, 143
Watanabe, T. 20, 21, 39, 40, 42, 43, 75, 77, 79, 95, 154, 155, 156, 190
Yamazaki, T. 198
Yoshioke, I. 149, 151
Zajaczkowska, A. 19, 40
Zimmer, K. 180

Subject Index

accommodation 9, 10, 15
alley
　parallel P- 18, 20-22, 30-33, 38, 40-42, 44-45, 61-64, 91-94, 190
　distance D- 18, 20-22, 30-33, 38, 40-42, 44-45, 64-65, 92-95, 96-97, 189
aniseikonia 165, 173
antipodal points 28
approach
　bottom-up 180-186
　top-down 197

basic
　circle BC 28, 55, 69
　sphere BS 55, 71
betweeness 180-182
binocular
　vision 8
　disparity 9, 16
body 2-3, 4

collinearity 23, 84-87, 181-184
conformal 26, 170
congruence 7, 189-193
constancy
　size 7, 15
　shape 7, 16
convergence 9, 10, 15, 33, 35, 36, 43, 126, 136, 144
corresponding points 9, 11-12
curvature
　center 47, 69
　Gaussian (total) K 23-24, 2-28, 40-46, 50-51, 147-148, 151-152. 155-156
　radius 47, 53, 69, 71
cyclopean vision 8-9, 11

Desarguesian 175-177
differential threshold 126
distance (perceived, visual) 10, 16, 25, 27, 55, 69, 80-84, 104-107, 137-143
　additivity 84-87
　radial 10, 27, 29, 35, 36, 104, 119-121, 122-133. 134-136, 144-145, 165, 196
　scaling 80-84
distorted room 17, 173

ego-centric localization 35-36, 144, 174
estimation of K and σ 37-39, 79, 94
Emmert's law 110
Euclidean map EM 26-29, 45, 52-56, 68-70, 107-109, 199
exocentric pointing 145-147

framework 4, 35
frameless 7, 17, 36, 45, 67, 149-163, 166, 168, 175-186, 198-199
Fréchet condition 23, 187
free-mobility 186, 190
frontoparallel
　curve H 19, 30-33, 38-40, 66, 70, 92-95, 147, 184
　plane HP 68-71, 95-97, 117, 170

Ganzfeld 5-6, 103, 113
geodesic 26-29, 48, 50, 69, 175

Helmholtz-Lie problem 186-189
Hillebrand hyperborae 165, 172, 199
horizon 5, 117-121

horopter
    horizontal 11, 19
    plane see frontoparallel plane
    vertical 11-14

iseikonic 36, 176
isometric 26
isotropic 36
induced movement 3

just-noticeable difference, jnd 12

Klein's model Σ 26, 53

line element 48-50, 57, 168-171, 175
Luneburg's mapping function 33-37, 38-39, 40-46, 67, 76, 78-79, 80, 87, 136, 149, 151, 153, 155, 158-160, 163-167, 171,173-175, 199

magnitude estimation 129, 130, 132, 133, 140, 143,147,
method of
    equal-appearing intervals 122, 135
    fractionation 133
    limits 150
multidimensional scaling MDS 80, 114, 137
    EMMDS 87-90, 92, 98, 137
    DMRPD 90-97, 94, 96-101, 114-117

origin O 9

perspective 40, 67, 193-197
physical space X 1, 24
plane, extending in the depth direction DP 9

horizontal DP 9
Poincaré model see Euclidian map

scale
    interval 133
    ratio 133
    based on difference judgment 122-129, 136
    based on ratio judgment 129-133, 136
self 1-4, 9
sigma σ 33, 36, 38, 42-43, 45, 75
similarity 7, 189-193
space (geometry)
    elementary 188
    elliptic 24-26, 52
    Euclidean 24-26, 52, 187
    Finsler 23
    G- 187-189
    hyperbolic 24-26, 52, 187
    locally Euclidean 23, 50, 143, 175, 189
    locally Minkowskian 23, 50, 143
    metric 23
    Riemannian 22-23
        constant curvature R 24, 26, 36, 53, 87, 107, 147, 151, 175-180, 186, 197-198
stimulus
    distant 1
    proximal 1, 6
subjective unit of distance sft, sm 123-126, 138-139

unit-sphere 50

vanishing point 175
Vieth-Müller circle 9, 11, 164-165, 171, 199
virtual reality 175

visual space VS 1, 24, 198
  boundary 4, 10, 28, 45-46, 119, 126
  features 2-7
  natural 103-148
  verdicality 5, 14, 35, 36, 145, 189

Credits

Permission to reproduce figures was obtained from the following copyright holders and is gratefully acknowledged.

American Medical Association, Chicago, Ill.
Parts of Chart2A and Chart3A in L.H.Hardy, G.Rand, and M.G. Rittler, Investigation of visual space, Archives of Ophthalmology, 1951, 45, 53-63, are reproduced in Fig.2.5.

Elsevier Ltd. Kidlingen, UK
Fig. 5 in T.Indow, An approach to geometry of visual space with no a priori mapping functions: Multidimensional mapping according to Riemannian metrics, Journal of Mathematical Psychology, 1982, 26, 204-236, is reproduced in Fig.2.6.
Fig.3 in J.Heller, On the psychophysics of binocular space perception, Journal of Mathematical Psychology, 1997, 41, 29-43, is used as the basis of Fig.5.6.

Pion Ltd. London, UK
Parts of Fig.1 and Fig.3 in T.Indow and T.Watanabe, Alleys on an extensive apparent frontoparallel plane: A second experiment, Perception, 1988, 17, 647-666, are reproduced in Fig.3.5.

American Psychological Association, Washington, DC.
Fig.13 in T.Indow, A critical Review of Luneburg's model with regard to global structure of visual space, Psychological Review, 1991, 98, 430-453, is reproduced in Fig.3.8.

Yale University Press, New Haven, CT.
Fig.80 in M.Kemp, The science of art, is used as the basis of Fig.5.22.